Bernadette Sharp and Rodolfo Delmonte (Eds.)
Natural Language Processing and Cognitive Science

De Gruyter Proceedings

Edited by
Bernadette Sharp
Rodolfo Delmonte

Natural Language Processing and Cognitive Science

Proceedings 2014

DE GRUYTER

Editors

Bernadette Sharp
Staffordshire University
Stafford, ST18 0AD
UK

Rodolfo Delmonte
Dorscoduro 175
30123 Venezia
Italy

ISBN 978-1-5015-1042-7
e-ISBN (PDF) 978-1-5015-0128-9
e-ISBN (EPUB) 978-1-5015-0131-9

Library of Congress Cataloging-in-Publication Data
A CIP catalog record for this book has been applied for at the Library of Congress.

Bibliographic information published by the Deutsche Nationalbibliothek
The Deutsche Nationalbibliothek lists this publication in the Deutsche Nationalbibliografie; detailed bibliographic data are available on the Internet at http://dnb.dnb.de.

Typesetting: PTP-Berlin, Protago TEX-Production GmbH
Printing and binding: CPI books GmbH, Leck
♾Printed on acid-free paper
Printed in Germany

www.degruyter.com

Preface

The 11th annual workshop of Natural Language Processing and Cognitive Science (NLPCS 2014) was held on 27–29 October, 2014 in Venice, hosted by the Department of Linguistic Studies, Ca' Bemardo (Ca' Foscari University). The aim of this workshop was to foster interactions among researchers and practitioners in Natural Language Processing (NLP) by taking a Cognitive Science perspective.

The workshop attracted 33 papers from 23 countries worldwide. Jared Bernstein, (Stanford University, USA) gave the keynote speech entitled "*Benchmarking Automated Text Correction Services*". The papers and posters presented at the workshops covered an impressive range of approaches ranging from linguistics, cognitive and computer science study to language processing.

We would like to thank the authors for providing the content of the programme. We are grateful to the programme committee who worked very hard in reviewing papers and providing feedback for authors. We would like also to thank Ca' Foscari University for hosting the workshop and for their help with housing and catering. Special thanks to Rocco Tripodi for his help with the administrative support of NLPCS website. Thanks in particular to de Gruyter for their support and help in publishing the proceedings.

We hope that you will find this program interesting and thought-provoking and that the workshop will provide you with a valuable opportunity to share ideas with other researchers and practitioners from institutions around the world.

October 2014

Co-chairs of the workshop:
Bernadette Sharp, Staffordshire University, U.K.
Rodolfo Delmonte, Ca' Foscari University, Italy

Contents

Workshop Chairs

Bernadette Sharp	(Staffordshire University, United Kingdom)
Rodolfo Delmonte	(Ca' Foscari University, Italy)

Programme Committee

Aretoulaki, M.	(Dialogconnection.com, UK)
Bosco, C.	(University of Torino, Italy)
Cabrio, E.	(INRIA Sophia-Antipolis Méditerranée, France)
Carl, M.	(Copenhagen Business School, Denmark)
Cristea, D.	(University A.I.Cuza, Iasi, Romania)
Day, C.	(Keele University, UK)
Delmonte, R.	(Universita' Ca' Foscari, Venice, Italy)
Endres-Niggemeyer, B.	(Fachhochschule Hanover, Germany)
Ferret, O.	(CEA, France)
Fisher, I.	(University of Konstanz, Germany)
Fuji, R.	(The University of Tokushima, Japan)
Higgins, S.	(Nottingham Trent University, UK)
Lenci, A.	(University of Pisa, Italy)
Lombardo V.	(University of Torino, Italy)
Magnini, B.	(Fondazione Bruno Kessler, Italy)
Mazzei, M.	(University of Torino, Italy)
Murray, W. R.	(Boeing Research and Technology, USA)
Neustein, A.	(International Journal of Speech Technology, USA)
Rapp, R.	(University of Aix-Marseille and Mainz, France/University of Mainz, Germany)
Rayson, R.	(Lancaster University, UK)
Roche, C.	(Condillac-LISTIC, Université de Savoie, France)
Schwab, D.	(LIG-GETALP, Grenoble, France)
Schwitter, R.	(Macquarie University, Sydney, Australia)
Sedes, F.	(Université de Toulouse, France)
Serraset, G	(Université Joseph Fourier – Grenoble 1, France)
Sharp, B.	(Staffordshire University, UK)
Thompson, G.	(Liverpool University, UK)
Tonelli, S.	(Fondazione Bruno Kessler, Italy)
Wandmacher, T.	(SYSTRAN, Paris, France)
Zock, M.	(LIF-CNRS, France)

Jared Bernstein, Alexei V. Ivanov, and Elizabeth Rosenfeld

Benchmarking Automated Text Correction Services

Abstract: We compared the performance of two automated grammar checkers on a small random sample of student texts to select the better text checker to incorporate in a product to tutor American students, age 12–16, in essay writing. We studied two competing error detection services that located and labeled errors with claimed accuracy well above "90%". However, performance measurement with reference to a small set of essays (double-annotated for text errors) found that both services operated at similar low accuracies (F1 values in the range of 0.25 to 0.35) when analyzing either sentence- or word-level errors. Several findings emerged. First, the two systems were quite uniform in the rate of missed errors (false negatives), but were very different in their distribution of false positives. Second, the magnitude of the difference in accuracy between these two text checkers is quite small – about one error, more or less, per two essays. Finally, we discuss contrasts between the functioning of these automated services in comparison to typical human approaches to grammar checking, and we propose bootstrap data collections that will support the development of improved future text correction methods.

1 Introduction

Student essays do not often satisfy all the doctrines of correctness that are taught in school or enforced by editors. Granting the use of "error" for deviation from prescribed usage, one observes that writers exhibit different patterns of error in text. There are several reasons why one would want to find, label, and fix those parts of text that do not conform to a standard, but our focus is on instruction. We recount a procedure we applied to compare the performance of two automated grammar checkers on a small random sample of student texts. The purpose was to select

Jared Bernstein: Stanford University, Stanford, California, USA, and Università Ca' Foscari, Venice, Italy, e-mail: Jared413@Stanford.edu
Alexei V. Ivanov: Fondazione Bruno Kessler, Trento, Italy, e-mail: Alexei_V_Ivanov@IEEE.org
Elizabeth Rosenfeld: Tasso Partners LLC, Palo Alto, California, USA, e-mail: Elizabeth@eRosenfeld.com

the better of two candidate text checkers to incorporate in a product designed to tutor American students, age 12–16, in essay writing.

Asking around, we found that experienced teachers approach text correction in several different ways, but teachers commonly infer the intent of the source passage and then produce a correct model re-write of a word or phrase that expresses that intent. For the struggling middle-school student, a teacher's analysis is often implicit and any instruction needs to be inferred or requested. By middle school, only some teachers categorize the errors and relate them to particular rules or rubrics as a specific means of instruction.

Burstein (2011) distinguishes between automated essay scoring (producing only an overall score) and essay evaluation systems, which provide diagnostic feedback. We'll follow her usage. Teachers can be trained to score essays with high ($r > 0.9$) inter-teacher reliabilities, and machines can be trained to match those teachers' essay scores with correlations at human-human levels. Scoring is easier than evaluation. Automated scoring can reach respectable levels of correlation with human scoring by combining a few proxy measures such as text length, text perplexity, and/or use of discourse markers, with the occurrence of word families extrapolated from a sample of human-scored texts on the same topic. Note that the score product is usually a single 3- or 4-bit value (e.g. A–F, or 1–10) that can be sufficiently optimized with regression techniques.

Essay evaluation, on the other hand, is more complicated for human teachers and for machines too. A 300-word essay by a typical middle school student (age 12–15) may have 10 or even 20 errors. Ideally, an evaluation will locate, delimit, label, and fix each error, and not label or fix other text material that should be accepted as correct or adequate. It is much more difficult to train teachers to evaluate and analyze a text than it is to train them in global essay scoring.

Let's make the over-simple assumption that a 300-word essay has 60 phrases and that there are 7 tag categories (1 correct and 6 error types) that we'd like to assign to each phrase. Also assume that about 75% of the phrases are "correct" and that the other six error categories are equally probable. If so, then just locating and labeling the errors in 60 phrases of a 300-word essay means producing about 86 bits of information. Note that delimiting the error (deciding which particular span of text is encompassed in a particular error) and selecting a correct replacement text is also part of what teachers routinely do, which creates more challenges for machine learning. Add the fact that teachers often approach an essay by only addressing the most egregious 5 or 10 errors, leaving relatively minor errors to be addressed in a later, revised draft from the student, and we see that a human-like scoring engine also needs to assign a severity score to each identified error. Therefore, full completion of the task probably needs to produce many more than 86 bits of information.

This differentiation of the text scoring and text evaluation tasks is parallel to the distinction between giving a second-language speaker an overall score on pronunciation (relatively easy) and reliably identifying the particular segmental and suprasegmental errors in a sample of that speaker's speech (quite hard). Even if one approaches the pronunciation scoring task as a simple combination of multiple error-prone individual evaluations of the segments and prosodic phrases, for the scoring task the many individual evaluation errors (false positive and false negative) are somewhat independent, so they can largely cancel each other out. In the evaluation task (as needed in a pronunciation tutoring system), the product is an evaluation of each unit (segment or phrase) so the errors add (as in a Word Error Rate calculation) and do not cancel each other out (see Bernstein, 2012).

This paper presents a comparison of two essay evaluation systems and gives some context relevant to that comparison. We present the results of an exploratory study of how teachers grade essays, compare typical teacher practices to the systems we studied, and finally propose a service model that might aggregate useful data for improving automated essay evaluation systems, while also helping teachers and students.

2 Prior Work on Essay Analysis

The field of automated essay analysis has generally been dominated by essay scoring systems, with a flurry of systems reported in the past 20 years, including ones by Burstein et al. (1998), Foltz, Kintsch, and Landauer (1998), Larkey (1998), Rudner (2002), and Elliott (2003). Page's (1966) Project Essay Grade (PEG) seems to be the first demonstration of essay scoring technology. PEG relied on counts of commas, prepositions and uncommon words, and unit counts for syllables, words and sentences. An early editing and proofreading tool was The Writer's Workbench developed by MacDonald et al. (1982), which gave feedback on points of grammar and orthographic conventions. The early scoring systems have typically implemented text feature extractors that feed a regression model that has been fit to match the average score of a panel of human essay scorers. Later, more sophisticated systems (Burstein et al., 1998) break an essay into sub-sections, homogeneous in topic domain, and extract similar features from those. Dikli (2006) presents an in-depth review of some major currently maintained systems.

Then, in the 1990s, scoring systems bifurcated, with some emphasizing form and convention and others focusing primarily on content. Landauer et al. (1998) proposed an ingenious text distance measurement called Latent Semantic Analysis (LSA) to evaluate the content of a written response and the appropriateness of vocabulary and collocation. By the late 1990's, machines had already attained correlations with human ratings on par with that observed between different human expert graders (~0.87–0.95) (Hearst, 2000).

Page and Petersen (1995), among others, criticized these grading systems for the indirect nature of features used for evaluation. The logic was that computers do not understand or appreciate the meaning of the assessed text, so the systems lack generality and are bound to topic training. Construction of the automated scorer for each new topic requires supervised retraining from a carefully verified human example. Automated graders only check some formal parameters that statistically correlate well with the human grade. Powers et al (2001) describe an ETS competition that demonstrated that the statistical nature of the scoring models is not robust to essays deliberately produced to fool an automated grading system. In fact, a rather casual and general observation of the decision criteria can inform the concoction of simple and effective methods to achieve high scores for low quality text. Shermis and Hamner (2013) describe a recent scoring competition on a sample of 22,000 middle school essays (most of which were hand written) and review the results from nine automated scoring engines. While confirming general criticism mentioned above they conclude that automated essay scoring is "capable of producing scores similar to human scores for extended-response writing items" and it "can be reliably applied in both low-stakes assessment ... and perhaps as a second scorer for high-stakes testing".

Another important point of criticism has been lack of constructive critical feedback to the essay author. Detection and classification of errors and infelicities has proved much more difficult. *What exactly is wrong? How can the writer make it better?* and *How can we teach a student to write more graceful and unambiguous prose?* These still seem to be intractable problems. Unlike the *classification problem* (i.e. separate and rank the overall-good from the overall-bad) of an essay scoring application, these questions ultimately generalize to the *detection problem* (i.e. point to the troublesome sub-parts of the text) and the *diagnosis problem* (i.e. determine exactly what is wrong here). On top of that, automated essay evaluation with analytical feedback has to operate on distinct, but interfering linguistic layers, e.g. discourse, argument, style, syntax, lexical usage, and, as always, orthographic conventions of format, spelling and punctuation.

More recent research has resulted in recipes to detect errors involving articles [Han et al., (2006)], prepositions [De Felice and Pulman (2008), Chodorow (2007), Tetreault & Chodorow (2008), Gamon et al. (2008), Gamon (2010), Han et al. (2010)], particles [Dickinson et al., (2011)], verb forms [Lee & Seneff (2008)], and collocations [Izumi et al., (2004), Dahlmeier & Ng, (2011)], grammar and usage errors [Rozovskaya & Roth (2010a, 2010b)]. These error-specific methods are reported to achieve precision up to 80–90% but they only capture specific, constrained error types such as misuse of prepositions and determiners. Dale & Kilgarriff, (2010) and Dale et al. (2012) describe a fairly complete system that aims to correct text written by English non-native speakers. These systems typically run a statistical decision making model over a window of successive token observations complemented with variable-complexity n-gram language model.

Error diagnosis seems idiosyncratic and resistant to human or machine convergence. There are often many valid ways to correct, improve or simplify a bad stretch of text (cf. Medero, 2014). In response, some universal methods of error detection, agnostic to the particular error type, have been proposed. Several statistical modeling techniques have been tried as universal error detection methods. Chodorow and Leacock (2000) tried mutual information and chi-square statistics to identify typical contexts for a small set of targeted words from a large well-formed corpus. Sun et al. (2007) went mining for patterns that consist of POS tags and function words. Wagner et al. (2007) used a combination of parse probabilities from a set of statistical parsers and POS tag n-gram probabilities, while Okanohara & Tsujii (2007) applied discriminative language models. Park & Levy (2011) report using a noisy channel model with a base language model and a set of error-specific noise models, and Gamon (2011) applied a maximum entropy Markov model to the problem, but no general solution has yet emerged.

One might apply automated text summarization techniques (e.g. Nenkova, (2012)) to measure the amount of useful information conveyed in an essay. Comparison of that with the original essay could be insightful on the efficiency of the author's text style. Intuitively, a good essay would gracefully provide large amounts of useful information in few words, while conforming to existing standards in its form. This approach may allow assessment of sentence specificity [Luis and Nenkova (2011)], general discourse structure [Luis and Nenkova (2013)].

The problem of essay correction (editing) can also be cast as a machine translation task. Indeed, we might consider the author's draft text as a source (with "draft English" as the separate language). In fact, it is the language with which the author attempts to communicate his ideas to the outside world. Then this source can be compared to the target language (i.e. standard English). A person's draft English might be biased by his L1 or by other cultural and social factors. Thus, any progress in the machine translation field [Wu (2009)] should be applicable to the essay correction task. The principal difficulty that exists here is lack of prior knowledge of the source language by the automated evaluation system. The system would generally have to infer it from the limited observable data in the essay to be evaluated and corrected.

3 Method

A publisher asked the authors to compare two competing error detection services (FB and QG) that located errors and provided feedback to students. Both FB and QG claimed accuracy well above 90%. We were told not to focus on meaning, style, coherence, or flow, because the publisher thought that its system already had these semantic and rhetorical aspects under control. We were instructed to focus on within-sentence issues of grammar, usage, conventions, spelling and the like. The assignment was to determine which engine, FB or QG, is more accurate and to estimate how large or significant the difference is between them.

It seems that the practice in the field may be to focus on the precision of the error location function, that is, the fraction of identified errors that are true errors. But we started out to find the F1 score for these evaluation systems. The problem with F1 and any other metric that counts false positive and false negative results is that "ground truth" must be known. A reference text with all the real errors marked is needed, which requires a careful and expensive process to locate and label errors in a sample of texts. As discussed above, such labeling is not simple and experts do not agree about what a good set of labels is nor about the right application of any given set of labels.

3.1 Materials

The materials used in the study are 24 essays, each approximately 300 words in length. An average essay in this set had about 18 sentences. The publisher's product is an essay-writing tutor that iteratively accepts draft texts and provides feedback for up to eight revisions of an essay. The data under study was obtained through random sampling of essays written by students aged 13 or 14, and from different stages in the iterative re-write process. The total sample comprised 440 sentences that contained 8840 word tokens.

Each essay was annotated (marked) independently by two highly literate readers. Disagreements between the first two readers were decided by a third reader. Each reader marked the essay texts according to a very specific set of rules. Each marked error was identified with one of five error types (SCUGL) defined as follows:

S: Bad sentence, too long, run-on, fragment, non-parallel, incoherent
G: Grammar, agreement, parallel verb form, ...
L: Spelling, includes capitalization, and special characters like ' - etc. within a word

U: Usage, includes wrong lexical item, or pronoun, or missing article, etc.
C: Convention, includes punctuation, spaces, non-standard abbreviation, etc.

The labels were inserted by hand into hundreds of sentence. So a sentence such as: "*... For example, only reasons countries has soldiers is because they fill they need them. ...*" results in an annotated text like:

... For example, «U» only «G» reasons countries «G»has soldiers «G» is because they («L»|«U») fill they need them. ...

In this hand marking scheme, a mark precedes the error and is bounded by « and ». In the example given above, wherever a mark is ineluctably ambiguous, as with the first and third «G», the mark is shown in gray type.

3.2 Procedure

Our evaluation of the automated marking systems proceeded in two stages: first, localization of the automatically detected problematic tokens matching those against a gold-standard reference text created by human experts; and second, a comparison of error type classification. Approximate string matching is essential for successful benchmarking of automated grammar correction. Localization was done using Levenshtein distance. The student's essay is initially pre-processed via tokenization to properly handle usage of punctuation marks. Unless the amount of differences in the proposed corrections is overwhelming, the Levenshtein algorithm proved to be robust and reliable in this task.

3.3 Analysis

Both essay-marking services always attempt error classification. Although the exact definitions of error classes are slightly different in the compared services, we tried to provide the most consistent mapping to our SCUGL categorization used in manual annotation. Error classification was evaluated through a confusion matrix between the automatically suggested and manually marked categories.

The final figures of merit for benchmarking were:

1. Sentence-level accuracy (percentage of true positive and true negative sentence-level errors);
2. token-level precision (the ratio of correctly identified grammatical problems to the number of all automatic detections);
3. token-level recall (the ratio of correctly identified grammar problems to the number of problems in the ground truth);

4. un-weighted token-level accuracy (share of confusion-matrix content on its main diagonal, normalized on the error class prior probabilities).

Chodorow et al. (2012) provide a helpful discussion of figures of merit for essay evaluation and conclude that “no single metric is best for all purposes”.

4 Results

Principal outcomes are summarized in the following tables. Table 1 presents the sentence-level correspondence between the ground-truth marking and each of the automatic systems (FB and QG).

Tab. 1: Sentence level performance of *FB* and *QG* a ground-truth reference; Analysis based on 440 sentences in 24 essays; T_+ – True Positive; T_- – True Negative; F_+ – False Positive; F_- – False Negative; Prec – Precision; Rec – Recall; *F1* – symmetric F-measure of accuracy.

System	T_+	F_+	T_-	F_-	Prec	Rec	*F1*
FB	15	16	370	39	0.48	0.28	0.35
QG	12	18	368	42	0.40	0.22	0.29

Table 2 presents similar measurements for the finer (token) level of detail. One can easily see that proportion of outcomes remains roughly similar between the two presented detail levels. Regardless the level of detail and taking into account the statistical properties of the experiment base (example set is quite limited), we can observe that the F1 value of both engines is around 0.3, which is very different from the claimed accuracy level. Comparing figures of merit in finer detail (Table 2) makes it evident that while the engines have similar values of F1, they are differently tuned: QG has better precision and FB is better in recall.

In both cases we have chosen to aggregate our findings with the symmetric accuracy F-measure:

$$F1 = 2 \times \frac{Prec \times Rec}{Prec + Rec}, \tag{1}$$

where precision and recall are defined in a classical manner (T_+, F_+ and F_- represent true and false positives and false negative experiment outcomes):

$$Prec = \frac{T_+}{T_+ + F_+}, \quad Rec = \frac{T_+}{T_+ + F_-}. \tag{2}$$

Although, arguably, it is a better practice to use a non-symmetric accuracy F-measure, e.g. one that favors precision:

$$F_\alpha = \frac{(1+\alpha^2)T_+}{(1+\alpha^2)T_+ + \alpha^2 F_- + F_+}, \quad \text{where } \alpha \in (0,1). \tag{3}$$

The value of α governs severity of the bias towards precision. As it can be easily shown, $\alpha = 1$ converts (3) into (1), while $\alpha = 0$ makes (3) to be equivalent to the definition of the precision. We shall also note that $\alpha \in (1,\infty)$ makes (3) to be biased favoring recall.

Tab. 2: Token level performance of *FB* and *QG* a ground-truth reference; analysis based on 8500 tokens in 440 sentences in 24 essays; T_+ – True Positive; T_- – True Negative; F_+ – False Positive; F_- – False Negative; Prec – Precision; Rec – Recall; $F1$ – symmetric F-measure of accuracy.

System	T_+	F_+	T_-	F_-	Prec	Rec	$F1$
FB	129	197	8021	493	0.40	0.21	0.27
QG	107	131	8061	514	0.45	0.17	0.25

Table 3 presents two confusion matrices, one for FB and one for QG. Each embedded matrix shows the accuracy of the labeling of text errors with reference to the ground-truth version of the marking. It is important to note that the confusion matrices were constructed for *correctly detected* token-level errors. This way the sum over each row in Table 3 gives a total number of correctly detected errors of a given type by the given engine. These figures are not obliged to sum to the same number for different engines.

Tab. 3: Confusion matrices for correctly detected token-level errors FB and QG a ground-truth reference; Error Types (*ET*): *G* – Grammar, *L* – spelLing, *U* – Usage, *C* – Convention, *O* – Other.

		Hypothesized *ET*				
System	**True *ET***	***G***	***L***	***U***	***C***	***O***
FB	*G*	**6**	0	2	2	0
	L	3	**19**	28	0	7
	U	0	1	**1**	0	0
	C	2	0	3	**55**	0
QG	*G*	**0**	0	2	1	–
	L	0	**30**	26	5	–
	U	1	0	**2**	0	–
	C	0	1	2	**37**	–

The FB service was attempting to give more detailed information regarding the nature of the detected error. However, for the purposes of the present comparative analysis the classes that do not fall into one of the defined error-type categories were summarized under "Other" label. It is surprising to see that, although spelling error *detection* is very well established with modern word-processors, spelling error *correction* task apparently is still more challenging. Both engines seemed to be randomly suggesting corrections through alternation of spelling or usage.

Figure 1 displays the distribution of false positives and false negative errors (while doing token-level detection of errors) by type for both the QG and the FB systems. Exessive number of G-, U-, C- and O-type errors jointly result in lower precision of the FB system.

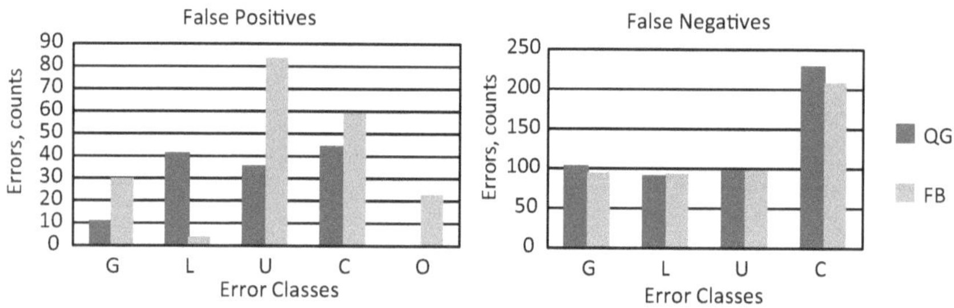

Fig. 1: Token-level error detection performance; Error Types: G – Grammar, L – spelLing, U – Usage, C – Convention, O – Other.

The results were clear for the publisher's purpose and can be summarized briefly as:

- Both engines have a high rate of error; F1 is about 0.3.
- Essay evaluation is difficult, context and author intent are unknown;
- With multiple ways to correct an error, choice often seems arbitrary;
- QG is better in precision (only finding real errors);
- FB is much better in recall (leaving fewer errors missed).

5 Discussion

The relative merits of bias towards precision or recall have system-wide implications. If one considers the automated correction service under analysis to be part of a complete system that's used to instruct real students,then precision may be more important than recall. Indeed, reporting obviously correct fragments (from the human perspective) as errors probably undermines student and teacher confidence in the quality of the feedback and could hamper the learning process. However, if the technologies (FB & QG) that we've measured are implemented in a system with a more intelligent post-processor (like a human teacher) that is able to undo false positive error marks, then not leaving any real errors unmarked (high level of recall) would be a greater a virtue. In this machine+human scenario, imperfections of the automatic evaluation would not *irreparably* damage the performance of the complete system.

5.1 Relation to Traditional Teacher Feedback

After the comparison of the two automated marking systems was completed, we started to think more generally about the problem of essay evaluation. It seemed to us that the most common method that teachers use to correct written assignments is to re-write the problematic parts. Whether the problem is limited to a single misspelled word or it's a confusing exposition of multiple actions spanning several sentences, we felt that teachers prefer to re-write. Re-writing is consistent with the pedagogy that advocates that learning is most effective if teachers model the correct behavior, rather than providing detailed criticism. To find out what teachers would actually do with the middle-school essays that we had machine-marked, we sent two of the 24 essays to 20 teachers to mark.

Written instructions to the teachers were simply this:

> Mark these TWO essays as if you are teaching a middle school English class. Make corrections/suggestions exactly as you would if you were going to hand this back to one of your students. You can track changes or attach comments in Word, or print this page and submit a picture of the edited essays.

Ten of the teachers responded – most with extensive markings and suggested rewrites. Eight of the ten teachers printed out the essays, marked them up with a pen, and then scanned or photographed their work. Close examination of these markings reveals a great diversity in approach. One teacher cares more about conveying the message clearly, another focuses on spelling and typographic conventions, while a third teacher wants to hone the style and usage.

We categorized the marked errors according to our 6-way error classification, and found that the teachers' markings were often much denser than the machine markings, but were quite inconsistent. That is, the teachers marked more errors (or made more suggestions) per word than the machines. The problem going forward is that the teachers' annotations are very inconsistent, both within teacher and especially across teachers. This can be compared again to the situation with an automated language tutor that attempts to identify specific phonetic segments that are mispronounced. Franco et al. (2010) reports that the simplest judgment (native-like or not) are quite variable at the segment level. Even among expert native listener-judges, inter-judge agreement is low with kappa values <0.40 for a majority of phonemes.

5.2 Proposal: Service in Support of Development

Given our current understanding of essay evaluation, one way forward would be to set up a service for teachers that collects essays and offers teachers a set of tools for correcting student draft texts. If the service had a convenient interface, presented feedback in a helpful form to the students, and aggregated statistics on error patterns for the teacher, teachers could be enticed to use it. If this service offered a uniform set of categories that constrained text alignments, then the resulting data would be a start on the path to collection of the materials we need to make substantial progress on an NLP problem that has so far resisted satisfactory solution.

Acknowledgements

Thanks Claire Neuner who ran the teacher evaluations and surveys, and coded the data.

Bibliography

Bernstein, J. (2012). Computer Scoring of Spoken Responses. *The Encyclopedia of Applied Linguistics*. J.Wiley. DOI: 10.1002/9781405198431.wbeal1044

Burstein, J., Kukich, K., Wolff, S., Lu, C., Chodorow, M., Braden-Harder, L., and Harris M. D. (1998). Automated Scoring Using A Hybrid Feature Identi?cation Technique. In: *Proceedings of the Thirty-Sixth Annual Meeting of the Association for Computational Linguistics*, East Stroudsburg, Penn.: Association for Computational Linguistics, 206–210.

Chodorow, M., and Leacock, C. (2000). An unsupervised method for detecting grammatical errors. In: *Proceedings of NAACL*, pp. 140–147.

Chodorow, M., Tetreault, J., and Han, N. R. (2007). Detection of grammatical errors involving prepositions. In: *Proceedings of the Fourth ACLSIGSEM Workshop on Prepositions*, pp. 25–30.

Chodorow, M., Dickinson, M., Israel, R., Tetreault, J. (2012). Problems in Evaluating Grammatical Error Detection Systems. In: Proceedings of COLING 2012, 611–628.

Dahlmeier, D. and Ng, H. T. (2012). Better evaluation for grammatical error correction. In: *Proceedings of NAACL-HLT*.

Dale, R. and Kilgarriff, A. (2010). Helping our own: Text massaging for computational linguistics as a new shared task. In: *International Conference on Natural Language Generation*.

Dale, R. and Narroway, G. (2012). A framework for evaluating text correction. In: Chair, N. C. C., Choukri, K., Declerck, S., editors, In: *Proceedings of the Eight International Conference on Language Resources and Evaluation (LREC'12)*, Istanbul, Turkey. European Language Resources Association (ELRA).

De Felice, R., and Pulman, S. G. (2008). A classifier-based approach to preposition and determiner error correction in L2 English. In: *Proceedings of COLING*, pp. 169–176.

Dickinson, M., Israel, R., and Lee, S. H. (2011). Developing methodology for Korean particle error detection. In: *Proceedings of the Sixth Workshop on Innovative Use of NLP for Building Educational Applications*, pp. 81–86, Portland, OR.

Dikli, S. (2006). An Overview of Automated Scoring of Essays. In: *The Journal of Technology, Learning, and Assessment*.Volume 5, Number 1 · August 2006

Elliott, S. (2003). Intellimetric: From Here to Validity. In: (ed.), *Automated Essay Scoring: A Cross-Disciplinary Perspective*. M. Shermis and J. Burstein. Hillsdale, N. J.: Lawrence Erlbaum Associates.

Franco, H., Bratt, H., Rossier, R., Rao, R., Shriberg, E., Abrash, V., & Precoda, K. (2010). EduSpeak: A speech recognition and pronunciation scoring toolkit for computer-aided language learning applications. *Language Testing*, 27(3), 401–18.

Foltz, P. W., Kintsch, W., and Landauer, T. K. (1998). *Analysis of Text Coherence Using Latent Semantic Analysis*. Discourse Processes 25(2–3): 285–307.

Han, N. R., Chodorow, M., and Leacock, C. (2006). Detecting errors in English article usage by non-native speakers. In: *Natural Language Engineering*, 12(2): 115–129.

Han, N. R., Tetreault, J., Lee, S.-H., and Ha, J. (2010). Using error-annotated ESL data to develop an ESL error correction system. In: *Proceedings of LREC*.

Hearst, M. A. (editor), (2000). *The debate on automated essay grading*, column, Trends & Controversies in IEEE Intelligent Systems, Sept.–Oct. 2000, pp. 22–37.

Izumi, E., Uchimoto, K., and Isahara, H. (2004). *SST speech corpus of Japanese learners' English and automatic detection of learners' errors*. International Computer Archive of Modern English Journal, 28:31–48.

Gamon. M., Gao, J., Brockett, C., Klementiev, A., Dolan, W., Belenko, D., and Vanderwende. L. (2008). Using Contextual Speller Techniques and Language Modeling for ESL Error Correction. In: *Proceedings of IJCNLP*.

Gamon. M., (2010). Using mostly native data to correct errors in learners' writing. In: Human Language Technologies: *The 2010 Annual Conference of the North American Chapter of the Association for Computational Linguistics (NAACL-HLT 2010)*, Los Angeles, California, pp. 163–171.

Gamon, M. (2011). High-order sequence modeling for language learner error detection. In: *Proceedings of the 6th Workshop on Innovative Use of NLP for Building Educational Applications*.

Landauer, T., Foltz, P. W., and Laham, D., (1998). Introduction to Latent Semantic Analysis. In: *Discourse Processes 25* (2–3), pp. 259–284.

Larkey, L. (1998). Automatic Essay Grading Using Text Categorization Techniques. In: *Proceedings of the Twenty-First ACM-SIGIR Conf. on Research and Development in Information*

Retrieval, New York, sociation for Computing Machinery Special Interest Group on Information Retrieval, pp. 90–95.
Lee, J. and Seneff, S. (2008). Correcting misuse of verb forms. In: *Proceedings of ACL-08: HLT*, Columbus, Ohio, pp. 174–182.
Louis, A., Nenkova, A. (2011). Automatic identi?cation of general and speci?c sentences by leveraging discourse annotations. In: *Proceedings of the 5th International Joint Conference on Natural Language Processing*, Chiang Mai, Thailand, November 8–13, 2011, pp. 605–613.
Louis, A., Nenkova, A. (2013). What Makes Writing Great? First Experiments on Article Quality Prediction in the Science Journalism Domain. *Transactions of the Association for Computational Linguistics*, 1 , 2013, 341–352. Action Editor: Mirella Lapata. Submitted 12/2012; Revised 3/2013, 5/2013; Published 7/2013.
MacDonald, N. H., Frase, L. T., Gingrich P. S., and Keenan, S. A. (1982). The Writer's Workbench: Computer Aids for Text Analysis. *IEEE Transactions on Communications 30*(1): 105–110.
Medero, J. (2014). *Automatic Characterization of Text Difficulty.* Dissertation: U. Washington.
Nenkova, A., McKeown, K. (2012). A Survey of Text Summarization Techniques. In: *Mining Text Data*, Springer US, 2012, pp. 43–76
Okanohara, D., and Tsujii. J. (2007). A discriminative language model with pseudo-negative samples. In: *Proceedings of ACL*, pp. 73–80.
Page, E. (1966). The Imminence of Grading Essays by Computer. *Phi Delta Kappan*, 48:238–243.
Page, E. B. & Petersen, N. S. (1995). The computer moves into essay grading: Updating the ancient test. *Phi Delta Kappan*, 76, 561–565.
Park, Y. A., and Levy, R. (2011). Automated whole sentence grammar correction using a Noisy Channel Model. In: *Proceedings of ACL* , 2011.
Powers, D. E., Burstein, J. C., Chodorow, M., Fowles, M. E., & Kukich, K. (2001). *Stumping e-rater: Challenging the validity of automated essay scoring (GRE Board Professional Rep. No. 98–08bP, ETS Research Rep. No. 01–03).* Princeton, NJ: Educational Testing Service.
Sun, G., Liu, X., Cong, G., Hou, M., Xiong, Z., Lee, J., and Lin, C. Y. (2007). Detecting erroneous sentences using automatically mined sequential patterns. In: *Proc. of ACL*, pp. 81–88.
Rozovskaya, A., and Roth, D. (2010a). Training Paradigms for correcting errors in grammar and usage. In: *Proceedings of NAACL-HLT.*
Rozovskaya, A., and Roth, D. (2010b). Generating confusion sets for context-sensitive error correction. In: *Proceedings of EMNLP.*
Rudner. L. M., and Liang, T. (2002). Automated Essay Scoring Using Bayes' Theorem. In: *Journal of Technology, Learning, and Assessment*, 1(2).
Shermis, M., and Burstein, J. eds. (2003). *Automated Essay Scoring: A Cross-Disciplinary Perspective.* Hillsdale, N. J.: Lawrence Erlbaum Associates.
Shermis, M. & Hamner, B. (2013). Contrasting State-of-the-Art Automated Scoring of Essays. In: M.Shermis & J.Burstein (eds.), *Handbook of automated essay evaluation: Current applications and new directions* (pp. 313–346). New York, NY: Routledge.
Tetreault, J. and Chodorow, M. (2008). The ups and downs of preposition error detection in ESL writing. In: *Proceedings of COLING-08*, Manchester.
Wagner, J., Foster, J., and van Genabith, J. (2007). Judging grammaticality: Experiments in sentence classification. In: *Proceedings of EMNLP & CONLL*, pp. 112–121
Wu, D. (2009). Toward machine translation with statistics and syntax and semantics. In: *ASRU 2009*, pp. 12–21.

Eniafe Festus Ayetiran, Guido Boella, Luigi Di Caro, and Livio Robaldo

Enhancing Word Sense Disambiguation Using A Hybrid Knowledge-Based Technique

Abstract: In this paper, we present a method for enhancing Word Sense Disambiguation through a hybrid system derived from the combination of Jiang & Conrath's similarity measure and a modified version of the original Lesk's algorithm by validating their results. The validation was carried out by computing the cumulative maximum overlap scores between the glosses of the senses returned by each of the algorithms and that of their semantically related senses in the WordNet lexical taxonomy with the original sentence containing the target word being disambiguated. Evaluation performed using the coarse-grained English all-words data of the SemEval 2007 showed that this method outperforms each of the two algorithms used alone and other independent state-of-the-art knowledge-based systems.

1 Introduction

Ambiguity is a fundamental characteristic of every language of which the English Language is not an exception. A considerable number of English words have more than one meaning. The meaning of word intended by a speaker or writer can be inferred considering the context of usage.

For example, consider the following sentences: (a) *My bank account yields a lot of interest annually* (b) *The children are playing on the bank of the river*. Based on the context of usage of the word "*bank*" in the two sentences above, we can infer that the first instance i.e sentence (a) is referring to a financial institution while the second instance, sentence (b) is referring to a sloping land beside a river. However, human identification of the right sense of a word as used in a statement is relatively easy but not with machines which need to process large unstructured textual information, carrying out complex computations in order to determine the right sense of the word as used in the statement.

Eniafe Festus Ayetiran: CIRSFID, University of Bologna, Italy, e-mail: eniafe.ayetiran2@unibo.it
Guido Boella, Luigi Di Caro, Livio Robaldo: Department of Computer Science, University of Turin, Italy, e-mail: {boella, dicaro, robaldo}@di.unito.it

The computational identification of the meaning of words in context is called Word Sense Disambiguation (WSD) also known as Lexical Disambiguation.

Basically, the output of any word sense disambiguation system is a set of words sense-tagged with the right synonymous word (if any). Considering the instances in the examples above, the sentences can be sense-tagged as follow: (a) *My* ***bank/financial institution/banking concern*** *account yields a lot of interest annually.* (b) *The children are playing on the* ***bank/sloppy land*** *of the river.*

Word Sense Disambiguation relies on knowledge. This means it uses a knowledge source or knowledge sources to associate the most appropriate senses with the words in context. Ideally, Word Sense Disambiguation is a means to an end but not usually the end itself, enhancing other tasks in different fields and application development such as parsing, semantic interpretation, machine translation, information retrieval and extraction, text mining, and lexical knowledge acquisition.

Approaches to word sense disambiguation may be knowledge-based (which depends on some knowledge dictionary or lexicon), supervised (machine learning techniques to train a system from labelled training sets) or unsupervised (based on unlabelled corpora, and do not exploit any manually sense-tagged corpus to provide a sense choice for a word in context). While knowledge-based and unsupervised approaches are automatic and do not involve manual training of systems, supervised approaches involve the herculean task of manually training systems with labelled datasets and often perform best on test sets from which part of the training sets have been taken.

The technique we used is a knowledge-based approach. We have employed English WordNet as the knowledge source and lexical resource. In the next section, we describe the English WordNet and discuss some relations in it that we have particularly employed for disambiguation in addition to those previously used in the literature. In the following sections, we also discuss related works, method, discussion of the results obtained and finally the conclusion.

2 WordNet

WordNet is a manually-constructed lexical system developed by George Miller (1990) and his colleagues at the Cognitive Science Laboratory at Princeton University and also made available in electronic form (Fellbaum, 1998). The basic object in WordNet is a set of strict synonyms called a synset. By definition, each synset in which a word appears is a different sense of that word. There are four main divisions in WordNet, one each for nouns, verbs, adjectives and adverbs.

Within a division, synsets are organised by the lexical relations defined on them. For nouns, the lexical relations include antonymy, hypernymy/hyponymy (IS-A relation) and three different meronymy/holonymy (PART-OF) relations. The verb also include the antonymy, hypernymy/hyponymy, troponymy and other relations like entailment, causes etc. The IS-A relation is the dominant relation, and organizes the synsets into a set of hierarchies.

We particularly focus on some of the important relations in WordNet which we have employed that had not been previously explored in word sense disambiguation. First is the derivationally related forms of a word which are generally thought to be highly regular and productive, and the addition of given affixes to their base forms produce new words whose meanings differ from that of the base words in a predictable way. Also used are the morphologically related nouns of adjectives and adverbs and this is applicable to adjectives and adverbs formed from nouns. For instance, the following can be derived from the word "*editorial*": "*editorialize*", "*editorialist*", "editor" and the noun form "voice" can be derived from the adjective "*vocal*". The second property is the antonymy relation which is the opposite of the target word e.g *wet* – *dry*. The third is entailment (verbs). According to WordNet glossary, entailment is defined as: "A verb X entails Y if X cannot be done unless Y is, or has been done" e.g *divorce entail marry*. The last is causes (verbs) and this can defined as an action or actions (causative) that triggers another action (resultative) e.g. *give* – *have*.

3 The Original Lesk's Algorithm

A basic knowledge-based approach relies on the calculation of the word overlap between the sense definitions of two or more target words. Micheal Lesk (1986) invented this approach named *gloss overlap* or the *Lesk's* algorithm. It is one of the first algorithms developed for the semantic disambiguation of all words in unrestricted texts. The only resource required by the algorithm is a set of dictionary entries, one for each possible word sense, and knowledge about the immediate context where the sense disambiguation is performed. The idea behind the Lesk's algorithm represents the starting seed for today's corpus-based algorithms. Almost every supervised WSD system relies one way or the other on some form of contextual overlap, with the overlap being typically measured between the context of an ambiguous word and contexts specific to various meanings of that word, as learned from previously annotated data.

The main idea behind the original definition of the algorithm is to disambiguate words by finding the overlap among their sense definitions. Namely, given two words, W_1 and W_2, each with N_{W_1} and N_{W_2} senses defined in a dictionary, for each possible sense pair W_1^i and W_2^j, $i = 1 \ldots N_{W_1}$, $j = 1 \ldots N_{W_2}$, we first determine the overlap of the corresponding definitions by counting the number of words they have in common. Next, the sense pair with the highest overlap is selected, and therefore a sense is assigned to each word in the initial word pair. The Algorithm is summarized in Listing 1 below:

```
1. for each sense i of W1
2. for each sense j of W2
3. compute overlap(i,j), the number of words in common
   between the definitions of sense i and sense j
4. find i and j for which overlap(i,j) is maximized
5. assign sense i to W1 and sense j to W2
```

Listing 1: The Original Lesk's Algorithm.

4 Related work

Since the seminal work of Michael Lesk (1986), several variations of the algorithm have been proposed. Each version of the algorithm's variation either (1) attempts to solve the combinatorial explosion of possible word sense combinations when more than two words are considered or (2) attempts to disambiguate, where each word in a given context is disambiguated individually by measuring the overlap between its corresponding dictionary definitions and the current sentential context and alternatives where the semantic space of a word meaning is augmented with definitions of semantically related words.

Cowie et al. (1992) worked on a variation called simulated Annealing. In this work, they define a function E that reflects the combination of word senses in a given text whose minimum should correspond to the correct choice of word senses. For a given combination of senses, all corresponding definitions from a dictionary are collected, and each word appearing at least once in these definitions receives a score equal to its number of occurrences. Adding all these scores together gives the *redundancy* of the text. The E function is then defined as the inverse of redundancy. The goal is to find a combination of senses that minimizes this function. To this end, an initial combination of senses is determined (e.g., pick the most frequent sense for each word), and then several iterations are performed, where the sense of a random word in the text is replaced with a different sense, and the new selection is considered as correct only if it reduces the value of the E function. The iterations stop when there is no change in the configuration of senses.

Banerjee & Pedersen (2002) developed a variation of the Lesk algorithm called the Adapted Lesk algorithm. Part of our technique builds on this, so we elaborate more on it than on other variations. Definitions of semantically related word senses in WordNet lexical hierarchy are used in addition to the definitions of the word senses themselves to determine the most likely sense for a word in a given context. Banerjee and Pedersen (2002) employed a function similar to the one defined by Cowie et al (1992) to determine a score for each possible combination of senses in a text and attempt to identify the sense configuration that leads to the highest score. While the original Lesk algorithm considers strictly the definition of a word meaning as a source of contextual information for a given sense, Banerjee and Pedersen extended this algorithm using WordNet. In their work, they employed WordNet synsets to obtain WordNet senses and their meanings (through their glosses). In addition, related concepts and the definitions of each word sense based on semantic lexical relations in WordNet were also used. These relations include hypernymy, hyponymy, meronymy etc. In their work, a limited window size of context words were used by considering only the words before and after the target word.

The algorithm takes as input an example or an instance in which target word occurs, and it will produce the sense for the word based on information about it and few immediately surrounding words. The crux of the work is the adaptation of WordNet semantic relations to the original Lesk's algorithm and computation of similarity based on maximum overlap of words (taking into account the number of times each word appears in the glosses).

5 Semantic Relatedness and Similarity

For a discourse to be coherent, words in the discourse must be related in meaning (Halliday and Hasan, 1976). This is a natural property of human languages and at the same time one of the most powerful constraints used in automatic word sense disambiguation. Words that share a common context are usually closely related in meaning (Harris, 1954), and therefore the appropriate senses can be selected by choosing those meanings found within the smallest semantic distance (Rada et al., 1989). While this kind of semantic constraint is often able to provide unity to an entire discourse, its scope has been usually limited to a small number of words found in the immediate vicinity of a target word, or to words connected by syntactic dependencies with the target word. These methods target the *local context* of a given word, and do not take into account additional contextual information found outside a certain window size. Similar to the Lesk's algorithm, these similarity methods become extremely computationally-intensive when more than two words are involved.

Patwardhan et al (2003) proposed that words in a discourse must be related in meaning for the discourse to be coherent. They distinguish between semantic relatedness and similarity following Budanitsky and Hirst (2004). They described semantic similarity as a kind of relatedness between two words that defines resemblance and semantic relatedness as a broader relationship between concepts that includes similarity as well as other relations such as is-a-kind-of, is-a-part-of as found in WordNet. Semantic similarity measures include Wu and Palmer (1994), Reisnik (1995), Agirre and Rigau (1996), Leacock et al. (1998), Hirst and St-Onge (1998), Lin (1998), Mihalcea and Moldovan (1999). Next, we discuss the Jiang and Conrath's measure which we used in our technique. We have adopted Jiang & Conrath's similarity measure for our final experiment because in our various experiments with each of the semantic similarity measures, we found it to be the best performing similarity measure.

6 Jiang & Conrath's Similarity Measure

Jiang & Conrath's similarity measure (Jaing & Conrath, 1998) is a similarity measure derived from corpus statistics and WordNet lexical taxonomy. It is a combined model that is derived from the edge-based notion by adding the information content as a decision factor. The model is based on the lexical taxonomy of the lexicon and statistics in the information content. In particular, attention is given to the determination of the link strength of an edge that links a parent node to a child node. Jiang & Conrath uses the difference in the information content of the two concepts to indicate their similarity. They used the information content defined by Resnik (1995) and augmented it with the notion of path length between concepts. This approach includes the information content of the concepts themselves along with the information content of their lowest common subsumer. A lowest common subsumer is a concept in a lexical taxonomy which has the shortest distance from the two concepts compared. The resulting formula can be expressed in Equation 1 below:

$$Similarity = \frac{1}{(\mathrm{IC(s1)} + \mathrm{IC(s2)} - 2 * \mathrm{IC(lcs)})} \tag{1}$$

Where *IC* is the information content, s_1 and s_2 are the first and second senses, respectively and *lcs* is the lowest common subsumer.

7 Method

In dealing with a raw text, we split the text into a set of sentences and the set of sentences split into a set of words called tokens. That is, for a text T, we split T into a set of finite sentences S_i, S ϵ T and tokenize each sentence S ϵ T into set of tokens W_i, W ϵ S. Each W ϵ S are then tagged into a part of speech. For monosemous words, the sense is returned as disambiguated based on the part of speech. For polysemous words, we followed the Adapted Lesk's approach of Bernajee and Pederson (2002) but we used the whole sentence as the window size. According to Heng Ji (2010), the antonymy relation is usually not exploited in WSD. However, if two context words are antonyms and belong to the same semantic cluster, they tend to represent the alternative attributes for the target word. Also, we opined that if two words are antonyms, there is possibility of mentioning the name of the opposite word in the gloss of the other opposite word, that of their semantically related senses or their examples and vice versa. Therefore, we also employed the *antonyms* in addition to hypernyms, hyponyms, meronyms etc used by Bernerjee and Pederson (2002). Also, for verbs we have used the *entailment* and *causes* relations. For adjectives and adverbs, we added the morphologically related nouns in computing the overlap score. In like manner, we disambiguated each target word in a sentence using the Jiang & Conrath's similarity measure using the whole sentence as the window size.

An agreement between the results produced by each of the two algorithms means the word under consideration has been rightly disambiguated and the sense on which they agreed is returned as the right sense. Wherever one method produces result and the other does not, the sense produced by the producing algorithm is returned as the right sense for the word being considered. Finally, in a situation where they produced different results, we validated the results returned by each of them. We did this by first obtaining the derivationally related forms of the words in context of the one under consideration and then add it to the original sentence to produce a new word list. Then for each of the senses produced by Modified Lesk and Jiang & Conrath algorithms, we obtain the maximum overlap between their glosses and that of their hypernyms, antonyms, hyponyms etc with the new word list. The sense with highest maximum overlap score is returned as the right sense. The intuition behind this notion of validation is that that the glosses of a word sense, and that of their semantically related ones in the WordNet lexical taxonomy should share words in common as much as possible with words in the original sentence. Adding the derivationally related forms of the words in the sentence increases the chances of overlap where there are changes in the word morphology. Finally, where the validation of senses could not arrive at a result due to lack of overlap, the result of the Modified Lesk is chosen as the right sense since it performs better than the Jiang & Conrath's similarity measure.

We illustrate the technique discussed above with the first sentence of the Semeval 2007 coarse-grained English all-words data. The target words in this sentence with their parts of speech are: *editorial – noun, ill – adjective, homeless – noun, refer – verb, research - noun, six – adjective, colleague – noun, report – verb, issue – noun.* In this case, we do not have to tokenize and tag into parts of speech since the data is a preprocessed one. The first word in the sentence which is a noun is monosemous because only a single noun sense exists for the word "*editorial*", therefore the single sense is returned as the right sense. The second word is "*ill*" and is an adjective. It is polysemous because the adjective form of it has five senses. To disambiguate using the Modified Lesk, for each of the senses of "*ill*", we obtained their glosses and that of their semantically related senses in the WordNet taxonomy. We also obtained the glosses and glosses of semantically related senses of "*editorial*", "*homeless*", "*refer*", "*research*", "*six*", "*colleague*", "*report*" and "*issue*" and computed the cumulative maximum overlap among them. The sense with the highest cumulative overlap score is returned as the right sense. To disambiguate the other words in the sentence, the same process was applied.

Furthermore, the Jiang & Conrath measure is not applicable to "*ill*" in this case because it requires the hierarchical relation of lowest common subsumer (lcs) in WordNet which adjectives and adverbs do not have. Using the next word "*homeless*" to illustrate this, for each of the two noun senses of "*homeless*", we compute the Jiang & Conrath similarity between each sense and that of each noun senses of "*editorial*", "*research*", "*colleague*", and "*issue*". This is because the lowest common subsumer can be measured for senses with the same part of speech. We also computed the cumulative similarity score and returned the sense with the highest score as the right sense of "*homeless*". Also, to disambiguate the other words in the sentence using Jiang & Conrath, the same process was applied.

Finally, the Modified Lesk's method returned sense "*homeless%1:14:00::*" as the right sense with a gloss of "someone without a housing" while the Jiang & Conrath method returned "*homeless%1:18:00*" with a gloss of "poor people who unfortunately do not have a home to live in". We added the derivationally related forms of each of the words in context of the target word to the original sentence and we have the following new word list: [*'editorial', 'editorialize', 'editorialist', 'editor', 'Ill', 'illness', 'Homeless', 'homeless', 'homeless', 'homeless', 'homelessness', 'homeless', 'refer', 'reference', 'reference', 'reference', 'referee', 'referee', 'reference', 'referral', 'reference', 'reference', 'reference', 'reference', 'research', 'research', 'research', 'research', 'researcher', 'research', 'researcher', 'research', 'research', 'six', 'colleague', 'collegial', 'report', 'report', 'report', 'report', 'report', 'report', 'report', 'report', 'report', 'report', 'report', 'reporter', 'report', 'report', 'reportage', 'reporting', 'issue', 'issue', 'issue', 'issue', 'issue', 'issue', 'issue', 'issue', 'issuer', 'issue', 'issuer', 'issuing', 'issue', 'issue', 'issue', 'issue'*].

Like we did for the Modified Lesk, we obtained the glosses each of "*homeless%1:18:00::*" and "*homeless%1;14;00::*", that of their hypernyms, hyponyms, meronyms, antonyms etc and computed the maximum overlap between them and the new word list. The final validation experiment returned "*homeless%1:18:00*" as the right sense. The same technique was applied for all the target words where the results of Modified Lesk and Jiang & Conrath algorithms do not agree.

8 Results and Discussion

In evaluating a Word Sense Disambiguation system, coverage is usually an important factor because it affects recall and F1 of the system. For systems that do not have a full coverage, back-off strategy is often used to supplement the results. It is a strategy where usually the Most Frequent Sense (MFS) of the target words are returned as the right sense wherever the system could not produce a result. However, the Most Frequent Sense (MFS) may not always help in all tasks as the accuracy varies from tasks to tasks and may not be ideal in practical application of Word Sense Disambiguation.

Our technique is strictly knowledge-based and independent as no back-off strategy or any form of supervision was employed. We evaluated our system with the coarse-grained English all-words data of the SemEval 2007 (Navigli et al., 2007) and presented the results obtained in Table 1 below:

Tab. 1: Analysis of Results by Part of Speech.

	Precision			
Method	**Noun**	**Verb**	**Adj**	**Adverb**
Modified Lesk	77.514	66.269	77.901	68.750
Jiang & conrath	76.748	66.263	-	-
KHybrid	82.581	68.020	-	-
	Recall			
Method	**Noun**	**Verb**	**Adj**	**Adverb**
Modified Lesk	76.534	65.820	77.901	68.750
Jiang & conrath	76.263	64.805	-	-
KHybrid	82.581	68.020	-	-
	F1			
Method	**Noun**	**Verb**	**Adj**	**Adverb**
Modified Lesk	77.044	66.043	77.901	68.750
Jiang & conrath	76.504	65.525	-	-
KHybrid	82.581	68.020	-	-

There are several knowledge-based and unsupervised word sense disambiguation systems in the literature but we pick a handful with the state-of-the-art results using the same evaluation dataset like our system. We present the final result of our hybrid technique in comparison with other knowledge-based/unsupervised systems using the same evaluation data in Table 2 below:

Tab. 2: Final Result and Results of other State-of-the-Art Knowledge-Based/Unsupervised Systems.

Method	Precision	Recall	F1
TKB-UO	70.207	70.207	70.207
Treematch	73.650	73.650	73.650
****KHybrid**	**76.509**	**76.509**	**76.509**
ExtLesk (with back off)	79.1	79.1	79.1
Degree (with back-off)	81.7	81.7	81.7

TKB-UO (Sánchez et al., 2007) was the best performing unsupervised system at the SemEval 2007 in the coarse-grained English all-words task. Treematch (Chen et al., 2009), ExtLesk (Ponzento & Navigli, 2012) and Degree (Ponzeto & Navigli, 2012) emerged after the workshop. Degree incorporates weak supervision in its implementation. KHybrid is our hybrid method. ExtLesk without back-off for nouns has precision of 82.7, recall of 69.2 and F1 of 75.4 while Degree without back-off for nouns has a precision of 87.3, recall of 72.7 and F1 of 79.4. The overall results of both ExtLesk and Degree without back-off were not made available.

9 Conclusion

Implementing both the Modified Lesk and Jiang & Conrath's algorithms using sentential window size are computationally intensive. However, they produce better results than using specific limited window sizes. Obtaining similarity between glosses of senses produced by each of the algorithms and the original sentences containing the target word under consideration improves the overall performance. In the future, we will try to experiment with this technique on several word sense disambiguation methods to evaluate and compare its effect on them.

Bibliography

Banerjee, S. & Pedersen, T. (2002) An adapted Lesk algorithm for Word Sense Disambiguation using WordNet. In: Gelbukh, A. (ed.), *Proceedings of the 3rd International Conference on Computational Linguistics and Intelligent Text Processing (CICLING)*, Mexico City, Mexico, 17–23 February, 2002, pp. 136–145

Cowie, J., Guthrie, J. & Guthrie, L. (1992) Lexical Disambiguation Using Simulated Annealing. In: Boitet, C. (ed.), *Proceedings of the 14th International Conference on Computational Linguistics (COLING)*, Nantes, France, 23–28 August, 1992, pp. 157–161

Miller, G. (1990) WordNet: An Online Lexical Database. *International Journal of Lexicography, 3(4): 235–244*

Jiang, J. J. & Conrath, D. W. (1998) Semantic similarity Based on Corpus Statistics and Lexical Taxonomy. In: Chen, K., Huang, C. and Sproat, R. (eds.), *Proceedings of the 10th International Conference on Research in Computational Linguistics*, Taipei, Taiwan, 2–4 August 1998, pp. 19–33

Lesk, M. (1986) Automatic Sense Disambiguation Using Machine Readable Dictionaries: How to Tell a Pine Cone from an Ice Cream Cone. In: Debuys, V. (ed.), *Proceedings of the 5th ACM-SIGDOC Conference*, Toronto, Canada, 8–11 June 1986, pp. 24–26

Heng J. (2010) One Sense per Context Cluster: Improving Word Sense Disambiguation Using Web-Scale Phrase Clustering. In: Zhou, A. and Nakamura, S. (eds.), *Proceedings of the 4th Universal Communication Symposium (IUCS)*, Beijing, China, 18–19 October 2010, pp. 181–184

Halliday, M. A. K. and Hasan, R. (1976) *Cohesion in English. English Language Series*. Longman, *New York*

Budanitsky, A., & Hirst, G. (2004) Evaluating WordNet-based measures of lexical semantic relatedness. *Computational Linguistics, 1(1): 1–49.*

Patwardhan, S., Banerjee, S. & Pedersen, T. (2003) Using Measures of Semantic Relatedness for Word Sense Disambiguation. In: Gelbukh, A. (ed.), *Proceedings of the 4th International Conference on Intelligent Text Processing and Computational Linguistics*, Mexico City, Mexico, 16–22 February 2003, pp. 241–257

Resnik, P. (1995) Using Information Content to Evaluate Semantic Similarity. In: Mellish, C. S. (ed.), *Proceedings of the 14th International Joint Conference on Artificial Intelligence (IJCAI)*, Montreal, Canada, 20–25 August 1995, pp. 448–453

Anaya- Sánchez, H., Pons-Porrata, A., & Berlanga-Llavori, R (2007) *TKB-UO:* Using sense clustering for Word Sense Disambiguation. In: Agirre, E., Màrquez, L., Wicentowski, R. (eds.), *Proceedings of the 4th International Workshop on Semantic Evaluations (SemEval)*, Prague, Czech Republic, 23–24 June 2007, pp. 322–325

Chen, P., Ding, W., Bowes, C. & Brown, D. (2009) A Fully Unsupervised Word Sense Disambiguation Method Using Dependency Knowledge. In: Doran, C., Ringger, E. (eds.), *Proceedings of the 10th Annual Conference of the North American Chapter of the ACL*, Boulder, Colorado, 31 May – 5 June 2009, pp. 28–36

Ponzeto, S. P. & Navigli, R. (2010) Knowledge-rich Word Sense Disambiguation Rivaling Supervised Systems. In: Hajic, J. (ed.), *Proceedings of the 48th Annual Meeting of the Association for Computational Linguistics*, Uppsala, Sweden, 11–16 July 2010, pp. 1522–1531.

Wu, Z. & Palmer, M. (1994) Verb semantics and lexical selection. In: Pustejovsky, J. (ed.), *Proceedings of 32nd Annual Meeting of the Association for Computational Linguistics,* Las Cruces, New Mexico 27–30 June 1994, pp. 133–138.

Navigli, R., Litkowski, K. C. & Hargraves, O (2007) *SemEval-2007 Task 07: Coarse-Grained English All-Words Task. In: Agirre, E., Màrquez, L., Wicentowski, R. (eds.), Proceedings of the 4th International Workshop on Semantic Evaluations,* Prague, 23–24 June 2007, pp. 30–35.

Lin, D. (1998) *An information theoretic definition of similarity. In: Jude, W. S (ed.), Proceedings of the 15th International Conference on Machine Learning,* Madison, U.S.A., 24–27 July 1998, pp. 296–304.

Hirst, G. & St-Onge, D. (1998) Lexical chains as representations of context in the detection and correction of malaproprisms. In: Fellbaum, C. (ed.), *WordNet: An Electronic Lexical Database,* pp. 305–332. MIT Press, Massachusetts, U.S.A.

Fellbaum, C. (ed.), (1998) *WordNet: An Electronic Lexical Database.* MIT Press.

Agirre, E. & Rigau, G. (1996) *Word sense disambiguation using conceptual density. In: Tsujii, J. (ed.), Proceedings of the 16th International Conference on Computational Linguistics (COLING),* Copenhagen, Denmark, pp. 16–22.

Mihalcea, R., & Moldovan, D. I. (1999) A method for word sense disambiguation of unrestricted text. In: Robert Dale and Ken Church (eds.), *Proceedings of the 37th annual meeting of the Association for Computational Linguistics on Computational Linguistics,* Maryland, USA, 20–26 June 1999, pp. 152–158.

Harris, Z. S. (1954) *Distributional Structure. Word,* 10(23): 146–162

Rada, R., Mili, H., Bicknell, E., & Blettner, M. (1989) Development and application of a metric on semantic nets. *IEEE Transactions on Systems,* Man and Cybernetics, 19(1): 17–30.

Franaçois Morlane-Hondère, Cécile Fabre, Nabil Hathout, and Ludovic Tanguy

Disambiguating Distributional Neighbors Using a Lexical Substitution Dataset

Abstract: This paper addresses the issue of polysemy in a distributional thesaurus. In such resources, distributional neighbors can relate indistinguishably to various senses. We propose a method to cluster the neighbors of a target word with respect to its senses, i.e. to attribute one sense to each neighbor. This is made possible by the use of a lexical substitution dataset, to which the distribution of the neighbors are compared.

1 Introduction

Many NLP applications need to know whether a word A is semantically more related to B than to C. Unsupervised corpus-based approaches to similarity have been widely used over the past 20 years. They are usually based on the *distributional hypothesis* (Harris, 1954): semantically related words tend to share many of their contexts (and vice versa). These approaches first collect information about the contexts in which words appear in a corpus, then measure the degree of relatedness between their distributions. The output is a *distributional thesaurus*, i.e. a list of word pairs – or *neighbors* – rated by a similarity score (Lin, 1998). *Distributional thesauri* have become very popular in a wide range of NLP tasks (Baroni and Lenci, 2010) because of the ever-increasing availability of textual data. However, their potential is still far from being fully exploited because of the fuzziness of the concept of *semantic relatedness* (Padó and Lapata, 2003).

In this paper, we address the issue of polysemy in distributional thesauri. For example, a polysemic word like *mouse* appears in different contexts according to its senses: *eat*_OBJ or *hunt*_OBJ in the ANIMAL sense and *plug*_OBJ or *cursor*_COMP in the DEVICE sense. These contexts are all mixed in the context vector of the word *mouse* when a large non-specific corpus is used. Therefore, the neighbors of *mouse* refer to both senses as in the following example taken from the Distributional Memory of Baroni and Lenci (2010), with the similarity score indicated for each

François Morlane-Hondère, Cécile Fabre, Nabil Hathout, Ludovic Tanguy: CLLE-ERSS: CNRS & University of Toulouse, France, e-mail: {francois.morlane, cecile.fabre, nabil.hathout, ludovic.tanguy}@univ-tlse2.fr

word (in decreasing order):

> *rat* (0.542), *animal* (0.466), *rabbit* (0.449), **cursor** (0.440), *monkey* (0.424), *cat* (0.421), *pig* (0.413), **joystick** (0.384), *dog* (0.375), *human* (0.373)...

The example shows that the ANIMAL sense is dominant in ukWaC (McCarthy et al., 2004) (the large web-based corpus from which the distributional data was extracted): it is referred to by eight out of the ten nearest neighbors of *mouse*; only *cursor* and *joystick* refer to the DEVICE sense.

Semantic relatedness can vary in both nature and strength. Wordnet-like resources focus on synonymy and hypernymy, but sense disambiguation can be extended to other semantic relations (such as, for the word *mouse*, meronymy: *fur* vs *cord*; co-hyponymy: *rat* vs *keyboard*) and general semantic relatedness (*cheese* vs *scrollbar*). We specifically address this question and show that disambiguation can be extended from standard synonymy to semantically more distant words using a distributional approach.

For that purpose, we designed an experiment whose setting is illustrated in figure 1. The objective is to cluster the neighbors of a target word with respect to its senses, i.e. to attribute one sense to each neighbor. For each sense of the target word we have a list of manually-validated similar words, more precisely words that have been proposed as suitable substitutes for the target word in different contexts (from the Semdis task, see section 3). The relatedness of a neighbor with a sense is thus estimated by the similarity of the former with the substitutes for the latter. The selection of the sense a neighbor is related to is performed in two steps:

Ⓐ The neighbor is compared to the substitutes of each of the target word senses.
Ⓑ The sense with the most similar substitutes is selected.

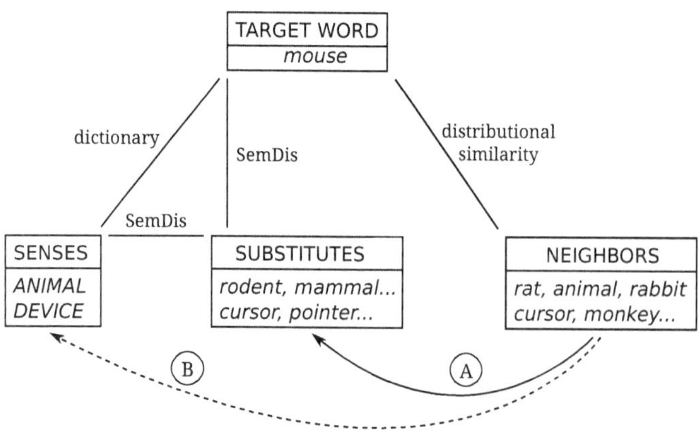

Fig. 1: Method overview.

The experiment described here has been performed on French, but can of course be easily ported to any language, provided that similar data are available.

The remainder of the paper is structured as follows: the task is described in Section 2. In Section 3, we focus on the sets of data we have used in this experiment. We then present the way we compute the similarity between the neighbors and the senses of the target words (Section 4). The details of the experiment are described in Section 5. We then give the results in Section 6.

2 The Task

Sorting out polysemy among pairs of distributional neighbors can be considered in relation to the word sense disambiguation (WSD) task (Agirre and Edmonds, 2006). The purpose of WSD is to identify the sense of a polysemous word used in a given context. Our approach differs from traditional WSD in that word disambiguation relies on its surrounding context as in (1), while what we want to do instead is to relate the distributional neighbors of a word to its different senses, as in (2).

(1) A little [mouse] is running around the room.

(2) *rat* (ANIMAL), *mammal* (ANIMAL), *rabbit* (ANIMAL), *cursor* (DEVICE)

WSD systems usually require two components. First, they need an inventory of the potential senses of the words. Most approaches rely on repositories such as WordNet (Fellbaum, 1998). However, when available, wordnets have been found to be too fine-grained to be useful (Ide and Véronis, 1998; Navigli, 2009), although efforts were made to make coarser the level of granularity of their sense distinctions (McCarthy, 2006). Then, most WSD systems need corpora. In many approaches, a sense-tagged corpus – like SemCor (Miller et al., 1994) – is used to train a classifier. Because no such a resource exists for French, we used an untagged corpus composed of French Wikipedia articles.

Our approach consists in computing a *second-order affinity* measure (Grefenstette, 1994) between the substitutes and the words for which we want to characterize the meaning. In the above example, we expect a higher similarity score between the vectors of *rabbit* and the substitutes for the sense ANIMAL than between *rabbit* and DEVICE. Karov and Edelman (1998) took a comparable approach, except that the meaning vectors – *feedback sets* – of their target words are made up of the words extracted from dictionary definitions. They call this method *similarity-based disambiguation.* For other examples of the use of second-order context vectors for WSD, see Pucci et al. (2009).

In Section 6.2 we compare this approach to a first-order (cooccurrence) affinity method. This approach is used by Yarowsky (1995) where the substitutes are seed words which the author considered representative of each target word sense. Both these approaches have been evaluated against a gold standard consisting in a manual annotation of each considered neighbor, for which the semantically closer meaning has been decided (see Section 5).

3 The Data

In this section, we present the datasets we have used in the experiment, namely a set of lexical substitutes grouped by senses, and a set of distributional neighbors for 11 target words.

3.1 The *SemDis 2014* Substitution Dataset

The dataset used in this study was created as a gold standard for a French lexical substitution task presented at the SemDis 2014 workshop (Fabre et al., 2014). This task was inspired by the SemEval 2007 English lexical substitution task (McCarthy and Navigli, 2009). Therefore, their goals and settings are quite similar: participants were asked to develop systems that give the best substitutes for a set of 30 target words – 10 nouns, 10 verbs and 10 adjectives – each contextualized in 10 sentences. The target words are polysemous words, chosen in a dictionary, and the sentences were picked in the FrWaC corpus (Baroni et al., 2009). Each sentence is associated to one of the word senses.The evaluation is performed by comparing the substitutes proposed by the participants' systems to a gold standard provided by the manual annotation of the sentences. The annotators – 7 different annotators for each sentence – were asked to produce single word substitutes for the target words in the sentence. From 0 to 3 substitutes could be given; for the 300 sentences, a total of 3961 substitutions were gathered (of which 1098 unique substitutes). The overall inter-annotator agreement was similar to what had been observed for this kind of task (25 % average pairwise agreement).

Below is an example of the resulting list of substitutes for the word *affection* 'affection', which may refer to either ILLNESS or LOVE. 5 sentences for each sense were proposed, and the corresponding substitutes were then merged, along with calculating the number of cases in which an annotator proposed each substitute.

– *affection* as ILLNESS: *maladie* (22), *trouble* (9), *pathologie* (7), *mal* (5), *dysfonctionnement* (3), *atteinte* (2), *problème* (2), *anomalie* (2), *condition* (1)...

- *affection* as LOVE: *amour* (26), *tendresse* (16), *attachement* (6), *amitié* (4), *sentiment* (4), *attention* (2), *lien* (2), *intimité* (1), *sollicitation* (1), *proximité* (1)...

This dataset is very limited in terms of coverage, but it has a number of advantages over more traditional dictionary- or wordnet-based data, as it fulfills the following requirements of our approach:
- each substitute is related to a sense;
- the semantic relatedness is valued (number of annotators);
- the substitutes are synonyms in a broader sense of the relation;
- all the substitutes are commonly used words and are thus frequent in contemporary corpora;
- the different senses are easily distinguished from one another.

To our knowledge, no resource exists that cumulates all these characteristics, which led us to select and use this dataset. The gold standard data is freely available from the SemDis workshop website[1].

In this study, we chose to first focus on a subset of 11 out of the 30 SemDis target words, those for which only two senses have been identified:
- 4 nouns: *affection* (LOVE VS ILLNESS), *débit* (DEBIT VS OUTPUT), *don* (TALENT VS PRESENT), *montée* (RISE VS ASCENT)
- 4 verbs: *entraîner* (TRAIN VS LEAD TO), *fonder* (CREATE VS BASE ON), *interpréter* (UNDERSTAND VS PERFORM), *maintenir* (HOLD VS ASSERT)
- 3 adjectives[2]: *aisé* (EASY VS RICH), *mince* (THIN VS POOR), *riche* (RICH VS WEALTHY)

3.2 The Distributional Thesaurus

We use a distributional thesaurus generated from a 262-million words corpus of French Wikipedia articles. The thesaurus contains the nouns, verbs and adjectives which occur in at least 5 different contexts in the corpus. Syntactic dependencies were used as contexts using the Talismane parser (Urieli and Tanguy, 2013). The weighting of the context relations was made using the pointwise mutual information and the cosine measure was used to compute the similarity between the context vectors.

1 http://www.irit.fr/semdis2014/fr/task1.html
2 There were only 3 two-senses adjectives in the 30 target words

Distributional information is used in two ways:
- it provides a list of its distributional neighbors for each of the 11 target words;
- it gives access to syntagmatic information in the Wikipedia corpus, corresponding to the syntactic contexts that characterize the substitutes and the neighbors (Section 2).

4 Method

In this section, we describe the way we compute the similarity between the neighbors and the senses of the target words. As mentioned in Section 2, for each target word, we compare the distribution of each neighbor with the overall distribution of the set of substitutes associated to each sense of this word.

So, what we first need to do is to build the context vectors of each neighbor as well as the context vectors of the 2 senses of each target word – or *sense vectors*. These are computed as follows:

1. we build a semantic space whose dimensions are the syntactic contexts of the target word in the Wikipedia corpus;
2. we compute the vectors of each substitute in this space, the values being the positive pointwise mutual information[3] (PPMI) between the substitute and a context;
3. we average the vectors of the substitutes of each sense to obtain a vector per sense. The weight for each substitute is the number of times it was proposed by the annotators, thus favoring the more common and natural substitutes.

In a second step, the cosine measure is used to compute the similarity between the vector of each neighbor and the two sense vectors. For example, we can see in table 1 that the correct sense systematically gets higher similarity scores (in bold).

5 Experiment

5.1 Gold Standard for Neighbor Disambiguation

To evaluate the ability of our method to disambiguate the target words, we performed a manual annotation of the 10 nearest neighbors for each of the 11 SemDis

3 A variant of the PMI where negative values are replaced by zero.

Tab. 1: Results of the similarity measure for the neighbors of the target word *affection*.

	senses	
	LOVE	ILLNESS
complication 'complication'	0.110	**0.590**
lésion 'lesion'	0.086	**0.744**
sympathie 'sympathy'	**0.680**	0.062
admiration 'admiration'	**0.776**	0.079
infection 'infection'	0.126	**0.691**
tumeur 'tumor'	0.052	**0.520**
symptôme 'symptom'	0.228	**0.542**
estime 'esteem'	**0.300**	0.024
manifestation 'manifestation'	0.134	**0.603**
épilepsie 'epilepsy'	0.083	**0.324**

target words we defined in Section 3. Neighbors that were also substitutes were excluded to avoid trivial configurations. For the purpose of this annotation, the two senses of each target word were arbitrarily assigned a number – *1* or *2* – and the most frequently proposed substitute was used as a gloss for each sense (like LOVE vs ILLNESS for the word *affection*).

In the annotation guidelines, we considered three options: a neighbor can be related to the first or the second sense of the target word (it will be annotated *1* or *2*), to both senses[4] (*0*) or to none (*–1*) when there is no identifiable semantic relation between the neighbor and the target word. Four judges have annotated all 110 neighbors. Overall inter-annotator agreement for this task, as measured by Fleiss' kappa, was 0.23, which traditionally indicates a "fair agreement". A majority vote was then applied to the four sets to produce a single annotation (only a few cases had to be decided through negotiation). Of the 110 neighbors, 77 were annotated *1* or *2*, 11 were annotated *0* and 22 were annotated *–1*.

5.2 Rule-Based Decision System

We designed a very simple rule-based decision system based on the similarity scores between the input neighbor and the two possible senses (see Section 4). Obviously, the higher similarity score should indicate the correct sense; however we also have the two other decisions to take into account. Our hypothesis is that

4 Mostly when the neighbor is morphologically related to the target word, like in *interpréter>interprétation*.

if the two similarity values are too close, it can indicate either a relation with both senses (if both values are high), either no relation (if both values are low).

We therefore need to base our decision on two variables, by comparing them to threshold values:

- the ratio between the two similarity values (higher score divided by lower): (threshold σ_r);
- the lowest of the two similarity values (threshold σ_{low}).

This rule is formalized in algorithm 1, where H is the sense for which the similarity value (sim_H) is higher and L (sim_L) the lower.

Algorithm 1 Decision rule:

```
if sim_H / sim_L > σ_r then
  output = sense H (1 or 2)
else
  if sim_L > σ_low then
    output = both senses (0)
  else
    output = no sense (−1)
  end if
end if
```

Both threshold values were estimated through a brute-force approach, and the highest accuracy was obtained with $\sigma_r = 1.08$ and $\sigma_{low} = 0.52$.

6 Results

6.1 Qualitative Analysis of the Results

The simple rule-based system presented in the previous section obtained an accuracy of 0.64 over the 110 neighbors. Details can be found in the confusion matrix (table 2). We can see that most of the errors are due to confusions with the categories *No sense* and *Both senses*.

When merging values for senses 1 and 2 (the distinction between the two categories being arbitrary) we reach a precision of 0.68, a recall of 0.84 and an f1-score of 0.75. Cohen's kappa value reaches 0.47, which indicates a much higher accuracy than what would be expected with random choices. As can be seen, there are only

Tab. 2: Confusion matrix for the rule-based system.

Gold \ System	No sense	Both senses	Sense 1	Sense 2	Total
No sense	5	0	2	15	22
Both senses	0	1	8	2	11
Sense 1	1	0	36	0	37
Sense 2	4	3	4	29	40
Total	10	4	50	46	110

4 cases out of 110 where the system chose one sense and the annotators chose the other.

We also found that the performance was rather unequal among the 11 target words. For example, the 10 neighbors of the word *affection* 'affection' were all related to their correct sense whereas this was only the case for 4 of the 10 neighbors of *fonder* 'found'. This is due to the fact that the two senses of the latter are more related than the senses of the former. This can be demonstrated by looking at the sense vectors of these two words. We reported in table 3 the 10 contexts with the

Tab. 3: Extracts of the meaning vectors of the target words *affection* and *fonder*.

***affection* 'affection'**		***fonder* 'found'**	
ILLNESS	**LOVE**	**CREATE**	**BASE ON**
neurologique_A 'neurologic'_A	*éprouver_V* 'feel'_V	*fait_N* 'fact'_N	*occasion_N* 'occasion'_N
système nerveux_NP 'nervous system'_NP	*particulier_A* 'particular'_A	*principe_N* 'principle'_N	*critère_N* 'criterion'_N
grave_A 'severe'_A	*profond_A* 'deep'_A	*hypothèse_N* 'hypothesis'_N	*guerre mondial_NP* 'world war'_NP
diagnostic_N 'diagnosis'_N	*grand_A* 'large'_A	*affirmation_N* 'statement'_N	*XIXe siècle_NP* 'XIXth century'_NP
souffrir_V 'suffer'_V	*vouer_V* 'vow'_V	*critère_N* 'criterion'_N	*remplacement_N* 'replacement'_N
rénal_A 'renal'_A	*témoigner_V* 'show'_V	*théorie_N* 'theory'_N	*emplacement_N* 'place'_N
mental_A 'mental'_A	*sentiment_N* 'feeling'_N	*idée_N* 'idea'_N	*modèle_N* 'model'_N
soigner_V 'cure'_V	*manifester_V* 'show'_V	*décision_N* 'decision'_N	*nom de institut_NP* 'institute name'_NP
chronique_A 'chronic'_A	*lien_N* 'link'_N	*témoignage_N* 'gesture'_N	*rive_N* 'bank'_N
oculaire_A 'ocular'_A	*paternel_A* 'paternal'_A	*raison_N* 'reason'_N	*même époque_NP* 'same era'_NP

higher PPMI for the two meaning vectors of the words *affection* and *fonder*. We can see that the contexts in which the substitutes related to the meaning ILLNESS appear are clearly related to the medical terminology, which strongly contrasts with the contexts related to the meaning LOVE. This distinction is far from being as clearly marked for *fonder*. We can see that most of the contexts of the meanings CREATE and BASE ON refer to abstract concepts (the context *critère* 'criterion' is even present in the two sense vectors).

6.2 Comparison with a First-Order Method

In order to get a better evaluation of our method, we designed a system in which second-order (distributional) similarity is replaced with first-order (cooccurrence). In other words, we used a similarity value between each neighbor and each substitute by calculating the PPMI based on their cooccurrence in a Wikipedia article (regardless of their frequency in the articles themselves). This PPMI score was then averaged as before to get a similarity score for both senses of the target words. This similarity values were then processed by the same rule-based system described above, with its own pair of optimal threshold values.

This technique reaches an overall accuracy of 50.91 %, and a f1-score of 0.60 for the "1 sense" target category. Both these scores are significantly lower than those obtained by our previous system.

7 Conclusion

The question of polysemy among the distributional neighbors addressed in this paper is part of a larger research effort concerning the description of the results provided by distributional semantics methods. In this experiment, we showed that a lexical substitution dataset can be effectively used to build distributional semantic vectors for distributional neighbors disambiguation. The use of a lexical substitution dataset was found to be an interesting alternative to the traditional wordnets and dictionaries, although more difficult to obtain in most cases. The idea of using the substitutes of a given sense to generate context vectors, and then to rely on these vectors to disambiguate the neighbors gives encouraging results considering the two modalities *0* and *–1*, usually absent from WSD tasks. Our method also sheds light on the fact that some sense vectors have more discrimination power than others. A cosine measure between the sense vectors of a given word could be helpful to assess its degree of polysemy (to merge the meanings of a fine-grained resource like WordNet, for example). In a future work, we plan to keep exploring the potential of our lexical substitution dataset for WSD in a less constrained framework, with a higher number of target words.

Bibliography

Agirre, E. and Edmonds, P. (2006). *Word Sense Disambiguation*. Springer.

Baroni, M., Bernardini, S., Ferraresi, A., and Zanchetta, E. (2009). The wacky wide web: a collection of very large linguistically processed web-crawled corpora. *Language Resources and Evaluation*, 43(3):209–226.

Baroni, M. and Lenci, A. (2010). Distributional memory: a general framework for corpus-based semantics. *Computational Linguistics*, 36(4):673–721.

Fabre, C., Hathout, N., Ho-Dac, L.-M., Morlane-Hondère, F., Muller, P., Sajous, F., Tanguy, L., and Van de Cruys, T. (2014). Présentation de l'atelier SemDis 2014 : sémantique distributionnelle pour la substitution lexicale et l'exploration de corpus spécialisés. In *Proceedings of the SemDis workshop*.

Fellbaum, C., editor (1998). *WordNet An Electronic Lexical Database*. MIT Press.

Grefenstette, G. (1994). *Corpus-derived First, Second, and Third-order Word Affinities*. Rank Xerox Research Centre.

Harris, Z. S. (1954). Distributional structure. *Word*, 10(23).

Ide, N. and Véronis, J. (1998). Word sense disambiguation: The state of the art. *Computational Linguistics*, 24(1).

Karov, Y. and Edelman, S. (1998). Similarity-based word sense disambiguation. *Computational Linguistics*, 24(1):41–59.

Lin, D. (1998). An information-theoretic definition of similarity. In *Proceedings of the Fifteenth International Conference on Machine Learning*, ICML '98, pages 296–304, San Francisco, CA, USA. Morgan Kaufmann Publishers Inc.

McCarthy, D. (2006). Relating wordnet senses for word sense disambiguation. In *Proceedings of the ACL Workshop on Making Sense of Sense*, pages 17–24.

McCarthy, D., Koeling, R., Weeds, J., and Carroll, J. A. (2004). Finding predominant word senses in untagged text. In *ACL*, pages 279–286.

McCarthy, D. and Navigli, R. (2009). The english lexical substitution task. *Language Resources and Evaluation*, 43(2):139–159.

Miller, G. A., Chodorow, M., Landes, S., Leacock, C., and Thomas, R. G. (1994). Using a semantic concordance for sense identification. In *Proceedings of HLT*, HLT '94, pages 240–243, Stroudsburg, PA, USA. Association for Computational Linguistics.

Navigli, R. (2009). Word sense disambiguation: A survey. *ACM Computing Surveys*, 41(2):10:1–10:69.

Padó, S. and Lapata, M. (2003). Constructing semantic space models from parsed corpora. In *proceeding of ACL*, pages 128–135, Sapporo, Japan.

Pucci, D., Baroni, M., Cutugno, F., and Lenci, A. (2009). Unsupervised lexical substitution with a word space model. In *Proceedings of EVALITA 2009*.

Urieli, A. and Tanguy, L. (2013). L'apport du faisceau dans l'analyse syntaxique en dépendances par transitions : études de cas avec l'analyseur Talismane. In *Actes de TALN*, pages 188–201, Les Sables d'Olonne, France.

Yarowsky, D. (1995). Unsupervised word sense disambiguation rivaling supervised methods. In *Proceedings of ACL*, pages 189–196, Stroudsburg, PA, USA.

Rocco Tripodi, Marcello Pelillo, and Rodolfo Delmonte

An Evolutionary Game Theoretic Approach to Word Sense Disambiguation

Abstract: This paper present a Semi Supervised approach to WSD, formulated in terms of Evolutionary Game Theory, where each word to be disambiguated is represented as a node on a graph and each sense as a class. The proposed algorithm performs a consistent class assignment of senses according to the similarity information of each word with the others, so that similar words are constrained to similar classes. The propagation of the information over the graph is formulated in terms of a non-cooperative multi-player game, where the players are the data points to decide their class memberships and equilibria correspond to consistent labeling of the data. The results have been provided by an analysis of their statistical significance. Preliminary experimental results demonstrate that our approach performs well compared with state-of-the-art algorithms, even though with the current setting it does not use contextual information or sense similarity information.

1 Introduction

Word Sense Disambiguation (WSD) is a task to identify the intended sense of a word in a computational manner based on the context in which it appears (Navigli, 2009). Understanding the ambiguity of natural languages is considered an AI-hard problem (Mallery, 1988). Computational problems like this are the central objectives of Artificial Intelligence (AI) and Natural Language Processing (NLP) because they aim at solving the epistemological question of how the mind works. It has been studied since the beginning of NLP (Weaver 1949) and also today is a central topic of this discipline.

Understanding the intended meaning of a word is a hard task, also humans have problem with it. This happens because people use words in a different manner from their literal meaning, and misinterpretation is a consequence of this

Rocco Tripodi, Marcello Pelillo: DAIS, Università Ca' Foscari di Venezia Via Torino 155, 30172 Venezia, Italy, e-mail: rocco.tripodi@unive.it
Rodolfo Delmonte: Language Science Department, Università Ca' Foscari, Ca' Bembo, dd. 1075, 30123, Venezia, Italy, e-mail: {pelillo, delmont}@unive.it

habit. In order to solve this task, it is not only required to have a deep knowledge of the language, but also to be an active speaker of the language. In addition, languages change over time, in an evolutionary process, by virtue of the use of language by speakers, and this leads to the formation of new words and new meanings which could be understood only be active speakers of the language.

WSD is also a central topic in applications like Text Entailment (Dagan and Glickman, 2004), Machine Translation (Vickrey, et al., 2005), Opinion Mining (Smrž, 2006) and Sentiment Analysis (Rentoumi et al. 2009). All of these applications require the disambiguation of ambiguous words, as preliminary process; otherwise they remain on the surface of the word (Pantel and Lin, 2002), compromising the coherence of the data to be analyzed.

The rest of this paper is structured as follows. Section 2 introduces the graph transduction technique and some notion of noncooperative game theory. Next, Section 3 shows our approach to WSD. Section 4 reports experimental results. Finally we review related studies in Section 5 and conclude in Section 6.

2 Graph-Based Semi-Supervised Learning

Our approach to WSD is based on two fundamental principles: the *homophily principle*, borrowed from social network analysis, and the *transductive learning*. The former simply states that objects which are similar to each other are expected to have the same label (Easley and Kleinberg, 2010). We extended this principle assuming that objects (words), which are similar, are expected to have a similar (semantic) class. The latter is a case of semi-supervised learning (Sammut and Webb, 2010) particularly suitable for relational data, which is used to propagate the class membership information from node to node. In our system we used an evolutionary process to propagate the information over a grah. This process is described in Section 2.2. In particular we extended the algorithm proposed by Erdem and Pelillo (2012) and used it in a scenario where the labels are fine-grained and their distribution is not uniform among the nodes of the graph. In fact in our case each node could be mapped to a disjoint set of classes (see Section 3).

2.1 Graph Transduction Model

Graph transduction is a semi-supervised method to solve labeling tasks wherein the objects to be labeled coincide with the nodes on a graph and the weights over the edges are the similarity measures among themselves. In this scenario, the geometry of the data is modeled with a weighted graph in which there are labeled and unlabeled points. The purpose of this method is to transfer the information given by the labeled points to the unlabeled ones, exploiting the pairwise similarity among nodes (*homophily principle*). Formally we have a graph $G = (D, E, w)$ in which D is the set of nodes representing both labeled and unlabeled points, $D = \{D_l, D_u\}$ and $w : E \rightarrow \mathbb{R}^+$ is a weight function assigning a similarity value to each edge $\epsilon \in E$. The task of transduction learning is to estimate the labels of the unlabeled points given the pairwise similarity for each node and a set of possible labels $\phi = \{1, \ldots, c\}$.

2.2 Normal Form Games

Game Theory provides predictive power in interactive decision situations. The simplest formulation of a game (normal form game, von Neumann and Morgenstern, 1943) assumes that games involve two players and p_{ij} is the payoff for player taking the ith action against player taking the jth action. The player's goal is to maximize his utility, which is the satisfaction a player derives from the outcome of the game computed by a utility function. In this kind of games it is assumed that the players are rational and have a complete knowledge of the strategy sets and associated payoffs (Szabó and Fath, 2007). An example of the information possessed by the players is described in Table 1, which depicts the famous prisoner dilemma game.

Tab. 1: Prisoner dilemma game.

p1/p2	Cooperate	Defect
Cooperate	1 , 1	0 , 2
Defect	2 , 0	0 , 0

As in Erdem and Pelillo (2012), we interpreted the graph transduction process in terms of noncooperative game, where each node[1] of the graph is a player ($i \in I$) participating in the game which can choose a strategy (class membership) among a set of strategies[2], $S_i = \{1, \ldots, c\}$ (classes). S_i is a probability distribution over its pure strategies which can be described as a vector $x_i = (x_{i1}, \ldots, x_{im_i})^T$ in which

1 In our case a node represents a target word in a co-occurrence network.

2 Target word senses.

each component x_{ih} denotes the probability that player i choose its hth pure strategy, where $\sum_{h=1}^{m_i} x_{ih} = 1$ and $x_{ih} \geq 0$ for all h.

In each game a player try to maximize its utility, choosing the strategy with the higher payoff. The final payoff of each player is given by the sum of payoff gained from each game played with its neighbors. Formally the pure strategy profile of player i is given by equation 1, where A_{ij} is the similarity value for player i and j.

$$\pi(s) = \sum_{j=1}^{n} A_{ij}(S_i, S_j), \tag{1}$$

This solution allows each player to exploit the information about the pure strategies of its neighbor and to converge to a consistent strategy, since the games are played only between similar players (neighbours) which are likely to have the same class and labeled players (or unambiguous words) do not play the game to maximize their payoffs but act as bias over the choices of unlabeled players (Erdem and Pelillo, 2012).

When equilibrium is reached, the label of a player is the strategy with the highest payoff, which is computed with equation 2.

$$\phi_i = \arg\max_{h=1,\ldots,c^{x_{ih}}} \tag{2}$$

3 WSD Games

3.1 Graph Construction

The graph is constructed selecting the target words from a dataset, denoted by $X = \{x_i\}_{i=1}^{N}$, where x_i corresponds to the i-th word to be disambiguated and N is the number of target words. From X we constructed the $N \times N$ similarity matrix W where each element w_{ij} is the similarity value assigned for the words i and j. W can be exploited as an useful tool for graph-based algorithms since it is treatable as weighted adjacency matrix of a weighted graph.

A crucial factor for the graph construction is the choice of the similarity measure, $s(\cdot,\cdot) \rightarrow \mathbb{R}$ to weights the edges of the graph. For our experiments we decided to use the the following formula to compute the word similarities:

$$w_{ij} = Dice(x_i, x_j) \forall i, j \in X : i \neq j \tag{3}$$

where $Dice(x_i, x_j)$ is the Dice coefficient (Dice, 1945) to determine the strength of co-occurrence between any two words x_i and x_j which is computed as follows:

$$Dice(x_i, x_j) = \frac{2c(x_i, x_j)}{c(x_i) + c(x_j)} \tag{4}$$

where $c(x_i)$ is the total number of occurrence of x_i in a large corpus and $c(x_i, x_j)$ is the co-occurrence of the words x_i and x_j in the same corpus. This formulation is particularly useful to decrease the ranking of words that tend to co-occur frequently with many other words. For our experiments we used as corpus the Google Web1T (Brants and Franz 2006), a large collection of n-grams (with a window of max 5 words) occurring in one terabyte of Web documents as collected by Google.

At this point we have the similarity graph W, which encodes the information of how two target words are similar, in a distributional semantics perspective (Harris, 1954). We recall that we will use W in order to classify similar words to similar classes, propagating the class membership information from node to node. For this reason, at first we need to smooth the data in W and then to choose only the most significant js for each $i \in X$. The first point is solved using a gaussian kernel on W, $w_{ij} = \exp \text{fi}(-\frac{w_{ij}^2}{2\sigma^2})$, where σ is the kernel width parameter; the second point is solved applying a $k$nearest neighbor (k-nn) algorithm to W, which allows us to remove the edges which are less significant for each $i \in X$. In our experiments we used $\sigma = 0.5$ and $k = 20$. Moreover, this operation reduces the computational cost of the algorithm and focuses only on relevant similarities.

3.2 Strategy Space

For each player i, its strategy profile is defined as:

$$S_i = \{s_{ij}\}_{j=1}^{c} \tag{5}$$

subject to the constraints described in Section 2.2, where c is the total number of possible senses, according to WordNet 3.0 (Miller, 1995). We can now define the strategy space S of the game in matrix form as:

$$\begin{matrix} s_{i1} & s_{i2} & \cdots & s_{ic} \\ \vdots & \vdots & \cdots & \vdots \\ s_{n1} & s_{n2} & \cdots & s_{nc} \end{matrix}$$

where each row correspond to the strategy space of a player and each column correspond to a class. Formally it is a c-dimensional space defined as:

$$\Delta_i = \{s_i \in \mathbb{R}^m : \sum_{h=1}^{m} s_{ih}, \text{ and } s_{ih} \geq 0 \text{ for all } h\} \tag{6}$$

each mixed strategy profile lives in the mixed strategy space of the game, given by the Cartesian product:

$$\Theta = \times_{i \in I} \Delta_i \tag{7}$$

For example in the case there are only two words, *area* and *country*, in our dataset, we first use WordNet in order to collect the sense inventories $M_i = 1, \ldots, m$ of each word, where m is the number of synsets associated to word i. Then we set all the sense inventories and obtain the set of all possible senses, $C = 1, \ldots, c$. In our example the two words have 6 and 5 synsets respectively, with a synset in common, so the strategy space S will have 10 dimension and could be represented as follows.

Tab. 2: strategy space.

S_{area}	$s_{1,1}$	$s_{1,2}$	$s_{1,3}$	$s_{1,4}$	$s_{1,5}$	$s_{1,6}$	$s_{1,7}$	$s_{1,8}$	$s_{1,9}$	$s_{1,10}$
$S_{country}$	$s_{2,1}$	$s_{2,2}$	$s_{2,3}$	$s_{2,4}$	$s_{2,5}$	$s_{2,6}$	$s_{2,7}$	$s_{2,8}$	$s_{2,9}$	$s_{2,10}$

At this point the strategy space can be initialized with the formula in order to follow the constraints described in Section 2.2.

$$s_{ij} = \begin{cases} |M|^{-1}, & \text{if sense } j \text{ is in } M_i. \\ 0, & \text{otherwise.} \end{cases} \tag{8}$$

The initialization of the strategy space is described in Table 3.

Tab. 3: strategy space initialization.

S_{area}	6^{-1}	6^{-1}	6^{-1}	6^{-1}	6^{-1}	6^{-1}	0	0	0	0
$S_{country}$	0	5^{-1}	0	0	0	0	5^{-1}	5^{-1}	5^{-1}	5^{-1}

In our example we have 10 synsets since *area* and *country* have a synset in common[3] (s_2), and each strategy profile S_i lies in the 10-dimensional simplex Δ_i. Each $s_{ij} \in S$ express a certain hypothesis about the membership of an object in a class, we set to zero senses, which are not in the sense inventory of a word to avoid that a word could choose a class, which is not in its intension.

3 s_2: (n) area, country (a particular geographical region of indefinite boundary, usually serving some special purpose or distinguished by its people or culture or geography) "it was a mountainous area"; "Bible country".

3.3 System Dynamics

Since we have the similarity graph W and the strategy space S, we can compute the payoff for each player according to equation (1). What is left is to compute the Nash equilibrium of the system, which means that each player's strategy is optimal against those of the others (Nash, 1951).

As in Erdem and Pelillo (2012) we used the dynamic interpretation of Nash equilibria in which the game is played repeatedly, until the system converges and dominant strategies emerge, updating S at time $t + 1$ with the following equation:

$$S_{ih}(t+1) = S_{ih}(t)\frac{u_i(e_i^h)}{u_i(s(t))} \tag{9}$$

which allows higher value of payoff to increase with higher rate, and where and where the utility function is computed as follows[4]:

$$u_i(e_i^h) = \sum_{j\in D_u} (A_{ij}, s_j)_h + \sum_{k=1}^{c} \sum_{J\in D_{l|k}} A_{ij}(h,k) \tag{10}$$

and

$$u_i(s) = \sum_{j\in D_u} s_i^t w_{ij} s_j + \sum_{k=1}^{c} \sum_{J\in D_{l|k}} s_i^t (A_{ij})_k \tag{11}$$

The payoffs associated with each player are additively separable and contribute to determine the final utility of each strategy $u_i(e_i^h)$.

4 Experimental Results

To make our experiments comparable to state-of-the-art systems, we used the official dataset of the SemEval-2 English all-words task (Agirre et al., 2010), which consists of three documents, 6000 word chunk with 1398 target words, annotated by hand using double-blind annotation by the organizer. Those texts were parsed using the TreeTagger tool (Schmid, 1994) to obtain lemmata and part-of-speech (POS).

The results are shown in Figure 1 where we compared our experiment with Semeval10 best, CFILT-2 (Khapra et al., 2010) (recall 0.57), and with the most common sense (MCS) approach (recall 0.505).

4 Note that A_{ij} is the partial payoff matrix for player i and j, computed multiplying the similarity weight w_{ij} with the identity matrix of size c, I_c.

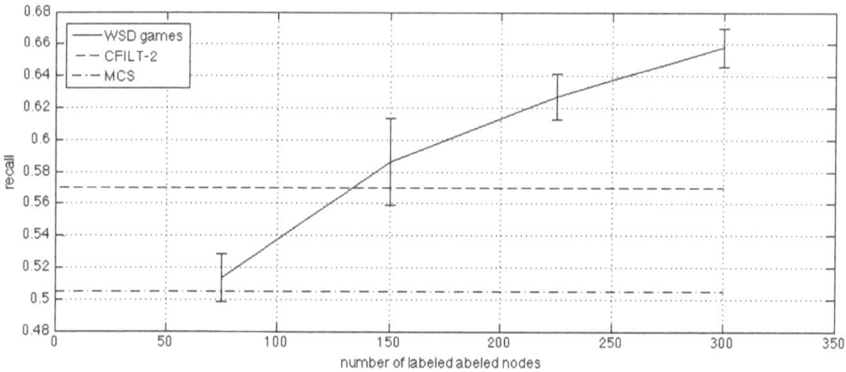

Fig. 1: Recall comparison

Figure 1 shows the mean recall rates and one standard deviation bars over 100 trials different sizes of randomly selected labeled points. We evaluated the results of our algorithm with the scoring tool provided with the emEval-2 English all-words dataset as described Agirre et al. (2010). Since our answers are in probabilistic format we evaluated all sense probabilities calculating the recall as the sum of the probabilities assigned by our system to the correct sense for each unlabeled word and dividing this sum by the number of unlabeled words, taking no notice of the labeled points which we used only as bias for our system.

The main difference between our approach and CFILT-2 is that we do not use domain specific knowledge-base in order to train our system. We need only a small amount of labeled data, which are easier to obtain and then propagate this information overt the graph. Our system is more similar to IIITTH, proposed by Agirre and Soroa (2009), which instead of using evolutionary dynamics over the graph, uses the personalized PageRank algorithm. This system has a recall of 0.534 on SemEval-2 English all-words task. Recently Tanigaki et al. (2013) proposed a smoothing model with a combinatorial optimization scheme for WSD, which exploits both contextual and sense similarity information to compute the maximum marginal likelihood for a sense given a specific target word. The main difference with this approach is semi-supervised, uses evolutionary dynamics in order to exploit the contextual information of a target word and with the current setting does not use contextual or sense similarity information, but only word similarity and relational information. This system has a recall of 50.8 on the same dataset we used for our experiments.

4.1 Conclusion

In this paper we presented a new graph based semi-supervised algorithm for WSD. Experimental results showed that our method improves the performances of conventional methods. Instead of train the system on large corpora, which are difficult to create and are not suitable for domain specific tasks, our system infers the class membership of a target word from a small amount of labeled data exploiting the relational information provided by labeled data. In our current implementation we use only the word similarity information in order to construct the graph. We think that the use of other sources of information could open new possible direction for further research in the area of transductive learning.

Bibliography

Agirre, E., Soroa, A.: Personalizing pagerank for word sense disambiguation. In: Proceedings of the 12th Conference of the European Chapter of the Association for Computational Linguistics, Association for Computational Linguistics (2009) 33–41

Agirre, E., De Lacalle, O.L., Fellbaum, C., Marchetti, A., Toral, A., Vossen, P.: Semeval-2010 task 17: All-words word sense disambiguation on a specific domain. In: Proceedings of the Workshop on Semantic Evaluations: Recent Achievements and Future Directions, Association for Computational Linguistics (2009) 123–128

Brants, T., Franz, A.: Web 1t 5-gram version 1. Linguistic Data Consortium, Philadelphia (2006)

Dagan, I., Glickman, O.: Probabilistic textual entailment: Generic applied modeling of language variability. (2004)

Dice, L.R.: Measures of the amount of ecologic association between species. Ecology **26**(3) (1945) 297–302

Easley, D., Kleinberg, J.: Networks, crowds, and markets. Cambridge University

Erdem, A., Pelillo, M.: Graph transduction as a noncooperative game. Neural Computation **24**(3) (2012) 700–723

Harris, Z.S.: Distributional structure. Word (1954)

Khapra, M.M., Shah, S., Kedia, P., Bhattacharyya, P.: Domain-specific word sense disambiguation combining corpus based and wordnet based parameters. In: In 5th International Conference on Global Wordnet (GWC2010, Citeseer (2010)

Mallery, J.C.: Thinking about foreign policy: Finding an appropriate role for artificially intelligent computers. In: Master's thesis, MIT Political Science Department, Citeseer (1988)

Miller, G.A.: Wordnet: a lexical database for english. Communications of the ACM **38**(11) (1995) 39–41

Nash, J.: Non-cooperative games. Annals of mathematics (1951) 286–295

Navigli, R.: Word sense disambiguation: A survey. ACM Computing Surveys (CSUR) **41**(2) (2009) 10

Pantel, P., Lin, D.: Discovering word senses from text. In: Proceedings of the eighth ACM SIGKDD international conference on Knowledge discovery and data mining, ACM (2002) 613–619

Rentoumi, V., Giannakopoulos, G., Karkaletsis, V., Vouros, G.A.: Sentiment analysis of figurative language using a word sense disambiguation approach. In: RANLP. (2009) 370–375
Sammut, C., Webb, G.I.: Encyclopedia of machine learning. Springer (2011)
Schmid, H.: Probabilistic part-of-speech tagging using decision trees. In: Proceedings of the international conference on new methods in language processing. Volume 12., Manchester, UK (1994) 44–49
Smrž, P.: Using wordnet for opinion mining. In: Proceedings of the Third International WordNet Conference, Masaryk University (2006) 333–335
Szabó, G., Fath, G.: Evolutionary games on graphs. Physics Reports **446**(4) (2007) 97–216
Tanigaki, K., Shiba, M., Munaka, T., Sagisaka, Y.: Density maximization in context-sense metric space for all-words wsd. In: ACL (1), Citeseer (2013) 884–893
Von Neumann, J., Morgenstern, O.: Theory of Games and Economic Behavior (60th Anniversary Commemorative Edition). Princeton university press (2007)
Vickrey, D., Biewald, L., Teyssier, M., Koller, D.: Word-sense disambiguation for machine translation. In: Proceedings of the conference on Human Language Technology and Empirical Methods in Natural Language Processing, Association for Computational Linguistics (2005) 771–778
Weaver, W.: Translation. Machine translation of languages **14** (1955) 15–23
Zhu, X., Ghahramani, Z.: Learning from labeled and unlabeled data with label propagation. Technical report, Technical Report CMU-CALD-02-107, Carnegie Mellon University (2002)

Merley S. Conrado, Thiago A. S. Pardo, and Solange O. Rezende

The Main Challenge of Semi-Automatic Term Extraction Methods

Abstract: Term extraction is the basis for many tasks such as building of taxonomies, ontologies and dictionaries, for translation, organization and retrieval of textual data. This paper studies the main challenge of semi-automatic term extraction methods, which is the difficulty to analyze the rank of candidates created by these methods. With the experimental evaluation performed in this work, it is possible to fairly compare a wide set of semi-automatic term extraction methods, which allows other future investigations. Additionally, we discovered which level of knowledge and threshold should be adopted for these methods in order to obtain good precision or F-measure. The results show there is not a unique method that is the best one for the three used corpora.

1 Introduction

Term extraction aims to identify a set of terminological units that best represent a specific domain corpus. Terms are fundamental in tasks for the building of (i) traditional lexicographical resources (such as glossaries and dictionaries) and (ii) computational resources (such as taxonomies and ontologies). Terms are also the basis for tasks such as information retrieval, summarisation, and text classification.

Traditionally, semi-automatic term extraction methods select candidate terms based on some linguistic knowledge [1]. After that, they apply measures or some combinations of measures (and/or heuristics) to form a rank of candidates [1–5]. Then, domain experts and/or terminologists analyze the rank in order to choose a threshold at which the candidates that have values above this threshold are selected as true terms. This analysis is subjective because it depends on personal human interpretation and domain knowledge, and it requires time to perform it.

Merley S. Conrado, Thiago A. S. Pardo, Solange O. Rezende: Instituto de Ciências Matemáticas e de Computação, Universidade de São Paulo, e-mail: {merleyc,taspardo,solange}@icmc.usp.br
Grants 2009/16142-3 Sao Paulo Research Foundation (FAPESP).

This subjectivity in analyzing the rank of candidates is the main challenge of semi-automatic term extraction methods.

Despite this challenge, the comparison of different extractors is a gap identified in the literature [6] since each research uses different corpora, preprocessing tools, and evaluation measures.

This paper aims to demonstrate how difficult it is to choose which candidates in a rank should be considered terms. For that, we perform and compare a wide set of term extraction methods that cover, separately, the three levels of knowledge used for term extraction: statistical, linguistic, and hybrid knowledge. We consider the same scenario when realizing the comparison of the extraction, i.e., we use the same corpora, the same textual preprocessing, and the same way of assessing results.

Our main contribution remains on demonstrate how difficult it is to analyze the rank of candidates created by semi-automatic term extraction methods. Additionally, in some cases, we discover which level of knowledge and threshold should be adopted for semi-automatic term extraction. Finally, with the experimental evaluation performed in this work, it is possible to fairly compare a wide set of semi-automatic term extraction methods, which allows other future investigations.

Next section describes the traditional term extraction methods and related work. Section 3 presents the measures used in the literature to extract terms, our experiments, results, and discussions. Finally, Section 4 presents conclusions and future work.

2 Related Work

Traditional term extraction select candidate terms based on some linguistic knowledge [1], e.g., to maintain only candidates that are nouns. Then, each candidate receives a value calculated by some statistical or hybrid measure or some combination of measures (and/or heuristics) [2–5]. These measures may be the candidate frequency or the accounting of distribution (e.g., the *weirdness* measure [7]) or occurrence probability (e.g., *glossEx* [3]) of candidates in a domain corpus and in a general language corpus. The candidates are ranked according to their values and those that have a minimum value of threshold in this rank are considered as potential terms of a specific domain. Domain experts and/or terminologists decide a threshold, which may be a fixed percentage or number of candidates to be considered. Manually choosing a threshold is not the best option since it has a high human cost. Semi-automatically choosing a threshold (e.g., an expert decision considers a fixed number of candidates) is not the best option as well, however it requires less human presence.

Luhn [8] and LuhnDF [9] are semi-automatic methods that plot histograms from candidate terms based on, respectively, candidate frequencies (*tf*) and document frequencies (*df*). These histograms facilitate the visualization of any possible pattern that candidates may follow and, then, the histograms help to determine a threshold. Salton et al. [10] propose another method that suggests to consider candidates that have *df* between 1% and 10% of the total number of documents in a corpus. The TRUCKS approach [11] suggests to consider only the 30% first candidates in the rank created according to the *nc-value* measure [2]. There are also studies that consider different fixed values of candidates [2, 12–14]. Other studies explore a variation of result combinations (precision and recall, usually) [5, 15]. There are studies that compare some measures used to extract terms [14, 16, 17].

All these studies use assorted ways to select a threshold in the candidate rank using different corpora and extracting, sometimes, simple terms and, other times, complex terms. This paper evaluates a wide set of simple term extraction methods comparing different thresholds and considering the same scenario of extraction. To the best of our knowledge, there is no reseaarch that evaluates this set of methods in a same scenario.

3 Evaluation of Traditional Term Extraction Methods

We performed and compared different semi-automatic extraction methods of simple terms. For the experiments, we use three corpora of different domains in the Portuguese language. The DE corpus [18] has 347 texts about distance education; the ECO corpus [13] contains 390 texts of the ecology domain; and the Nanoscience and Nanotechnology (N&N) corpus [19] has 1,057 texts.

In order to minimize the interference of the preprocessing in the term extraction, we carefully preprocessed all the texts as follows:

1. We identify the encoding of each document in order to correctly read the words. Without this identification, we would incorrectly find "*p*" and "*s-teste*" instead of "*pós-teste*" (*post-test*). We also transformed all letters to lowercase.

2. We remove stopwords[1] and special characters, such as \, |, , and ^.
3. We clean the texts, such as to convert "*humanos1*" to "*humanos*" (*humans*), "*que_a*" to "*que*" (*that*) and "*a*" (*an*), and "*tam-be'm*" to "*também*" (*too*). In these examples, *humans* would be a candidate term and the other words might be removed (because they are stopwords).
4. We identify part-of-speech (*pos*) tags using the Palavras parser [20].
5. We normalize the words using stemming[2].
6. As domain experts, we also consider compound terms (e.g., "*bate-papo*" – *chat*) as simple terms.

At the end of preprocessing, we obtained 9,997, 16,013, and 41,335 stemmed candidates, respectively, for the ECO, DE, and N&N corpora.

3.1 Term Extraction Methods

Each preprocessed unigram of each corpus was considered a candidate term. For all candidates of each corpus, we evaluated, separately, 21 simple term extraction methods. These methods are divided into three levels of knowledge: statistical, linguistic, and hybrid knowledge.

Each statistical method applies some statistical measure (Table 1) in order to quantify termhood, i.e., to express how much a candidate is related to the corpus domain. In Table 1, D is the number of documents in a corpus (c); f_{d_x,t_j} is the frequency of t_j (j^{th} candidate term) in the d_x (x^{th} document); $1 - p(0; \lambda_j)$ is the Poisson probability of a document with at least one occurrence; and W is the amount of corpus words.

We used the *tv*, *tvq*, and *tc* measures – normally applied to the attribute selection tasks (identified by *) – because they were considered relevant measures for extracting terms in the work of [26]. We used the *n-gram length* measure to verify if terms of a specific domain have different length (in characters) of words in general language or of terms in another domain. E.g., the longest term of the ecology domain ("*territorialidade*" – *territoriality*) contains 16 characters, while the longest term of the N&N domain ("*hidroxipropilmetilcelulose*" – *hydroxypropylmethylcellulose*) has 26 characters.

1 Stoplist and Indicative Phrase list are available at http://sites.labic.icmc.usp.br/merleyc/ThesisData/

2 PTStemmer: a stemming toolkit for the Portuguese language – http://code.google.com/p/ptstemmer/

Tab. 1: The statistical measures.

Acronym	Measure	Equation
1. *n-gram length*	Number of characters in a *n*-gram	–
2. *tf*	Term frequency	$\sum_{x=1}^{D} f_{d_x,t_j}$
3. *rf*	Relative frequency	tf_{t_j}/W
4. *atf*	Average term frequency	tf_{t_j}/df_{t_j}
5. *ridf*	Residual inverse document frequency [21]	$\left(\log_2\left(\frac{D}{df_{t_j}}\right)\right) - \log_2\left(\frac{1}{1-p(0;\lambda_j)}\right)$
6. *df*	Document frequency	$\sum_{x=1}^{D}(1 \mid f_{d_x,t_j} \neq 0)$
7. *tf-idf*	Term frequency – inverse document frequency [22]	$tf_{d_x,t_j} \times \log\left(\frac{D}{df_{t_j}}\right)$
8. *tv**	Term variance [23]	$\sum_{x=1}^{D}[f_{d_x,t_j} - \bar{f}_{t_j}]^2$
9. *tvq**	Term variance quality [24]	$\sum_{x=1}^{D} f_{d_x,t_j}^2 - \frac{1}{D}\left[\sum_{x=1}^{D} f_{d_x,t_j}\right]^2$
10. *tc**	Term contribution [25]	$\sum_{x=1}^{D}\sum_{y=1}^{D} f_{d_x,t_j} \times idf_{t_j} \times f_{d_x,t_j} \times idf_{t_j}$

We expect that the statistical measures *tf*, *rf*, *atf*[3], *ridf*[3], and *tf-idf* help to identify frequent domain terms. The *df* measure counts in how many documents in the corpus the candidate terms occur. Then, we expect that *df* identifies candidates that represent the corpus by assuming they occur in at least a minimal amount of documents.

There are frequent terms in the corpus, but there are also rare terms or those that have the same frequency of non-terms. The statistical measures are not able to identify these differences. For this reason, we also evaluated four linguistic methods of extracting terms that follow different ways for obtaining linguistic knowledge aiming to identify terms. We used the annotation provided by the

3 We used the implementation available at https://code.google.com/p/jatetoolkit/

Palavras parser [20]. The first linguistic method of extracting terms considers terms are *noun phrases*. The second linguistic method (*pos*) assumes terms are nouns. The third linguistic method (*k_noun_phrase*) considers terms are kernels of noun phrases, since they represent the meaningful kernels of terms, as discussed in [27]. For instance, the noun phrase "*Os autótrofos terrestres*" (*The terrestrial autotrophics*) belongs to the ecology domain and the experts of this domain consider only the kernel of this phrase as a term, i.e., *autotrophics*. It is also expected that, if the texts' authors define or describe some word, the latter is important for the text domain and, for this reason, this word is possibly a term. For example, in "*A segunda possibilidade de coalescência é descrita por...*" (*The second possibility of coalescing is described by*), the term *coalescence* from the nanoscience and nanotechnology domain may be identified because it is near the indicative phrase *is described by*. Thus, the fourth linguist extraction method (*ip*) considers terms are those that occur near some indicative phrase.

When considering only statistical measures, it is not possible to identify, e.g., terms with similar frequencies to non-terms. Similarly, when considering only linguistic measures, it is not possible to identify terms that follow the same patterns of non-terms. Then, we expect that to consider the statistical and linguistic knowledge together may optimize the term identification. For example, the verb *to propose* can be quite frequent in technical texts from a specific domain. Although, if we assume a term should follow both a linguistic pattern (e.g., being a noun) and a statistical pattern (e.g., being frequent in the corpus), this verb – even if it is frequent – will be correctly identified as non-term.

For this reason, we also evaluated six hybrid methods of extracting terms found in the literature. Two of these methods use statistical and linguistic knowledge to identify terms by applying, separately, the *c-value* and *nc-value* measures (Table 2). The other hybrid methods statistically analyze information of general language corpus by applying, separately, the *gc_freq.*, *weirdness*[3], *thd*, *tds*, and *glossEx*[3] measures (Table 2). The latter methods assume, in general, that terms have very low frequencies or that do not appear in general language corpus. We used the NILC[4] corpus of general language with 40 million words. For the identification of phrases used in the *c-value* and *nc-value* measures and to find kernels of noun phrases and part-of-speech tags, we used the annotation provided by the Palavras parser [20].

In Table 2, $r_{t_j}^{(c)}$ is the ordination value of the candidate t_j in a specific domain corpus c; g is a general language corpus; $td(t_j)$ is domain specificity of t_j; and $tc(t_j)$ is the cohesion of the t_j candidate. For *c-value*, T_{t_j} is the candidate set with length

4 NILC Corpus – http://www.nilc.icmc.usp.br/nilc/tools/corpora.htm

Tab. 2: The hybrid measures.

Acronym	Measure	Equation
1. *gc_freq.*	Term frequency in a general language corpus	$\sum_{x=1}^{D^{(g)}} f_{d_x,t_j}$
2. *weirdness*	Term distribution in a domain corpus and general language corpus [7]	$\left(tf_{t_j}^{(c)}/W^{(c)}\right) / \left(tf_{t_j}^{(g)}/W^{(g)}\right)$
3. *thd*	Termhood index: weighted term frequency in a domain corpus and general language corpus [4]	$\left(r_{t_j}^{(c)}/W^{(c)}\right) - \left(r_{t_j}^{(g)}/W^{(g)}\right)$
4. *tds*	Term domain specificity [28]	$\left(P(t_j{}^{(c)})/P(t_j{}^{(g)})\right) = \frac{\textit{prob. in domain c}}{\textit{prob. in corpus g}}$
5. *glossEx*	Occurrence probability of a term in a domain corpus and general language corpus [3]	$a * td(t_j) + b * tc(t_j)$, default a=0.9, b = 0.1.
6. *c-value*	Frequency of a andidate with certain *pos* in the domain corpus	$(1 + \log_2 \lvert t_j \rvert) \times \log_2 \lvert t_j \rvert \times tf(t_j)$, if $t_j \notin$ a V; otherwise
	and its frequency inside other longer candidate terms [2, 29]	$(1 + \log_2 \lvert t_j \rvert) \times \log_2 \lvert t_j \rvert \left(tf(t_j) - \frac{1}{P(T_{t_j})} \sum_{b \in T} f(b)\right)$.
7. *nc-value*	The context in which the candidate occurs is relevant [2]	$0.8\ \textit{c-value}_{t_j} + 0.2 \sum_{b \in C_{t_j}} f_{t_j}(b)(t(w)/nc)$

in grams larger than t_j and that contains t_j; $P(T_{t_j})$ is the number of such candidates (types) including the type of t_j; and V is the set of neighbours of t_j. For *nc-value*, C_{t_j} is the set of words in the context of the candidate t_j; b is a context word for the candidate t_j; $f_{t_j}(b)$ is the occurrence frequency of b as a context word for t_j; w is the calculated weight for b as a context word; and nc is the total number of candidates considered in the corpus.

3.2 Results and Discussion

For the term extractors that use statistical and hybrid measures (Tables 1 and 2), which result in continuous values, the candidates are decreasingly ordered by

Tab. 3: Extraction Method that uses *tf*: The ECO corpus.

#Cand.	#ET	P(%)	R(%)	FM(%)
50	21	42,00	7,00	12,00
100	32	32,00	10,67	16,00
150	45	30,00	15,00	20,00
200	52	26,00	17,33	20,80
250	61	24,40	20,33	22,18
300	69	23,00	23,00	23,00
350	77	22,00	25,67	23,69
400	85	21,25	28,33	24,29
		...		
9800	298	3,04	99,33	5,90
9850	298	3,03	99,33	5,87
9900	298	3,01	99,33	5,84
9950	299	3,01	99,67	5,83

their chance of being terms, considering the value obtained in the calculation of each measure. In this way, those candidates that have the best values in accordance with a certain measure are at the top of the rank. Then, we calculated precision, recall, and F-measure considering different cutoff points in this rank, starting with the first 50 ordered candidates, the first 100 candidates, 150, and so on until the total number of candidates of each corpus. To calculate precision, recall, and f-measure, we used gold standards of the ECO, DE, and N&N corpora, which contain, respectively, 322, 118, and 1,794 simple terms, and, after stemming them, we have 300, 112, and 1,543 stemmed terms. The authors of the ECO corpus built its gold standard with the unigrams that occur, at the same time, in 2 books, 2 specialized glossaries, 1 online dictionary, all related to the ecology domain. Differently, an expert of the distance education domain decided which noun phrases, that satisfied certain conditions, should be considered terms. For the elaboration of the N&N gold standard, its authors applied statistical methods to selected candidate terms, then manually a linguist removed some of them and, finally, an expert decided which of these candidates are terms.

Table 3 shows the results[5] of extraction method that uses the *tf* measure with the ECO corpus. This table also highlights the highest precision (P (%) = 42.00), recall (R (%) = 99.67), and F-measure (FM (%) = 24.29) achieved when using the *tf*

5 All the term extraction results for the three corpora using each measure are available at http://sites.labic.icmc.usp.br/merleyc/ThesisData/.

measure with, respectively, the first 50, 400, and 9,950 candidates of the rank (# Cand.).

The linguistic methods result a unique and fixed total number of extracted candidates with which we calculated the evaluation measures. For example, all the 5,030 noun phrases identified in the ECO corpus are considered extracted candidates. Considering there are 279 terms (#ET) in these candidates, 5.55% of the extracted candidates are terms (P(%)), identified 93% of the total of terms in this domain (R(%)), and achieved a balance between precision and recall equal to 10.47% (FM(%)).

Table 4 shows the best and worst results[5] for precision, recall, and F-measure of the term extraction methods using different measures considering various cut-off points in the candidate rank.

We observe that the best precision (52% – line 1 in Table 4) of the ECO corpus was achieved when using the top 50 candidates ordered by the *tc* measure. The best recall (100% – line 7) was reached with *gc_freq.*, *weirdness*, and *thd* when using more than 88% (8,850 candidates) of the corpus. The best F-measure (29.43% – line 13) used the top 400 candidates ordered by *tvq*. Regarding the DE corpus, the best precision (36% – line 2) was performed when using the first 50 candidates better ranked by the *tds* measure. The best recall (100% – line 8) was reached using *tf, rf, atf, df, tf-idf, tv, tvq, tc, gc_freq.*, and *c-value* considering more than 11,305 candidates (> 70% of the DE corpus). On the other hand, the best F-measure (22.22% – line 14) was achieved using the top 50 candidates ordered by the *tds* measure. Finally, for the N&N corpus, the best precision (66% – line 3) was obtained using the top 50 candidates ranked by the *tvq* measure. The best recall (94.10% – line 9) was achieved with *tf* and *rf* using more than 14,900 candidates (> 36% of the corpus). The best F-measure (36.22% – line 15) was reached with the first 3,150 candidates ordered by *tf-idf.*

Figure 1 shows the relation between the number of extracted candidates and values of precision, recall, and F-measure obtained with these candidates considering the ECO corpus. Due to the limited page number of this paper, we only show the graphic of the ECO corpus, however, we discuss the results of the three used corpora. For these three corpora, the highest precisions are achieved using the same amount of candidates, which is 50. The recall values reach around 100%, however, most of these cases use almost the entire corpus. Accordingly, the recall values are not considered good results since, if the entire corpus is used, the results would be equal or similar (see also lines 19–21 in Table 4). There are methods that achieve around 20% to 30% of F-measure for the ECO and DE corpora when using from 0.31% to 4,5% of these corpora (from 50 to 450 candidates). Meanwhile, for the N&N corpus, it is necessary to use 7.62% of the corpus (3150 candidates) to obtain the best F-measure (36.22%).

Tab. 4: Summary of term extraction method results.

	Line	Corpora	Measures	#Cand.	#ET	P (%)	R (%)	FM (%)
Best	1	ECO	tc	50	26	52,00	8,67	14,86
precision	2	DE	tds, glossEx	50	18	36,00	16,07	22,22
	3	N&N	tvq	50	33	66,00	2,14	4,14
Worst	4	ECO	n-gram length, tf, rf, atf, df, tf-idf, tv, tvq, tc, c-value	9950	299	3,01	99,67	5,83
precision	5	DE	ridf	15000	64	0,43	57,14	0,85
	6	N&N	ridf	14950	16	0,11	1,04	0,19
Best	7	ECO	gc_freq., weirdness, thd, nc-value	> 8850	300	3,02 - 3,39	100,00	5,85 - 6,56
recall	8	DE	tf, rf, atf, df, tc, tf-idf, tv, tvq, gc_freq., c-value	> 11100	112	0,75 - 1,01	100,00	1,48 - 2,00
	9	N&N	tf, rf	> 14900	1452	9,68 - 9,74	94,10	17,55 - 17,66
	10	ECO	tds, glossEx	50	2 - 4	4,00	0,67 - 1,33	1,14 - 2,00
Best	11	DE	thd	50	6	12,00	5,36	7,41
recall	12	N&N	weirdness	50	16	0,11	1,04	0,19
	13	ECO	ridf tvq	14950 400	103	25,75	34,33	29,43
Best	14	DE	tds, glossEx	50	18	36,00	16,07	22,22
F-measure	15	N&N	tf-idf	3150	850	26,98	55,09	36,22
	16	ECO	tds, glossEx	50	2 - 4	4,00	0,67 - 1,33	1,14 - 2,00
Worst	17	DE	ridf	15000	64	0,43	57,14	0,85
F-measure	18	N&N	ridf	14950	16	0,11	1,04	0,19
	19	ECO	–	9950	300	3,02	100,00	5,85
Entire	20	DE	–	16000	112	0,70	100,00	1,39
corpus	21	N&N	–	41335	1543	0,04	100,00	0,07

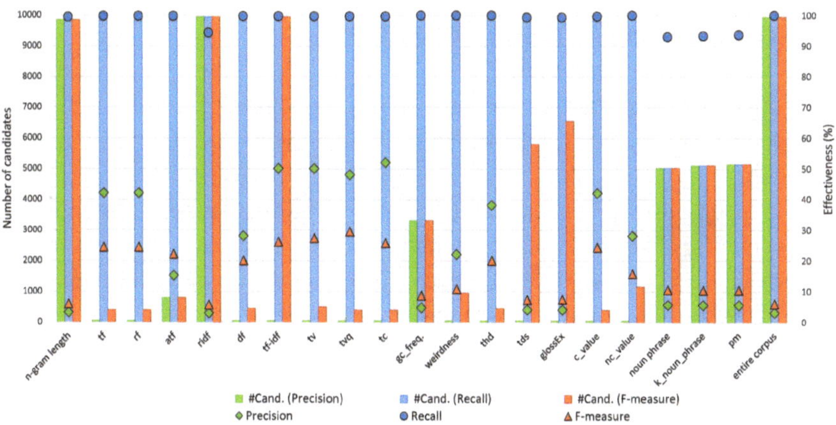

Fig. 1: Precision, recall, and F-measure *vs.* amount of extracted candidates – The ECO corpus.

Regarding the knowledge level (linguistic, statistical, or hybrid) of the methods, in general, the highest precision results were obtained using the statistical methods or some hybrid methods (such as *tds* or *c-value*). Interestingly, those methods that use measures do not commonly used to extract terms (*tc* or *tvq*), were a good option considering their precision values. The recall results of the statistical and hybrid methods were very similar because their best recall results use almost 100% of the candidates; this fact makes the recall independent of the used measures. An exception of the statement about the independence of recall compared to the measures is the use of the linguistic measures, since the extractors that use them have the highest recall values: 93.67%, 96.43%, and 88.32% using 51%, 49%, and 42% of the candidates, respectively, for the ECO, DE, and N&N corpora.

When using statistical or hybrid extractors, F-measure generally maintains the same pattern. It is noteworthy that the linguistic extractors obtain low precision and F-measure results (between 1.35 to 8.09%). This fact give us evidence the linguistic extractors should use measures of other knowledge levels as well. The linguistic extractors achieve high recall results (between 85 and 96%), which means they are able to remove part of the non-terms without excluding many terms. Therefore, we proved that better results are obtained when the statistical measures are applied on a candidate list that was previously filtered based on some linguistic measure, as stated by [1].

4 Conclusions and Future Work

With the experiments performed in this work, we demonstrated how difficult it is to analyze the rank of candidates created by semi-automatic term extraction methods.

Based on our experiments, we list below some suggestions to be followed by the traditional semi-automatic methods of simple term extraction. In order to achieve good precisions, we suggest to consider the first 50 candidates ordered by some of the statistical or hybrid methods. It was not possible, however, to identify which method is the best one to reach good recall, since the highest recall values were only achieved using (almost) the entire corpus, which is not recommended. Linguistic methods showed to be promising for that. It was not possible to identify a unique method that is the best one for the three used corpora. Nevertheless, regarding the threshold used in the candidate ranks, the statistical (except *n-gram length* and *ridf*) and hybrid (except *gc_freq.*) methods are the most desirable when aiming to achieve high precision and F-measure.

Regarding the four extractors that use measures (*tv*, *tvq*, *tc*, and *n-gram length*) normally applied to other tasks instead of term extraction, we observe that *tv*, *tvq*, and *tc* were responsible for at least one of the highest results of each corpus. Therefore, as expected, these three measures are good options for extracting terms. However, *n-gram length* reached lower results than the other measures used in this research. Then, we conclude that, contrary to expectations, there is no difference in the length (in characters) between terms and non-terms of these corpora. Finally, this experiments demonstrate how difficult and subjective it is to determine a threshold in the candidate term ranking. Our future work remains on combining the measures and exploring new ones.

Bibliography

[1] Pazienza, M.T., Pennacchiotti, M., Zanzotto, F.M.: Terminology extraction: An analysis of linguistic and statistical approaches. In Sirmakessis, S., ed.: Knowledge Mining Series: Studies in Fuzziness and Soft Computing. Springer Verlag (2005) 255–279

[2] Frantzi, K.T., Ananiadou, S., Tsujii, J.I.: The C-value/NC-value method of automatic recognition for multi-word terms. In: PROC of the 2nd European CNF on Research and Advanced Technology for Digital Libraries (ECDL), London, UK, Springer-Verlag (1998) 585–604

[3] Kozakov, L., Park, Y., Fin, T.H., Drissi, Y., Doganata, Y.N., , Confino, T.: Glossary extraction and knowledge in large organisations via semantic Web technologies. In: PROC of the 6th INT Semantic Web CNF (ISWC) and the 2nd Asian Semantic Web CNF. (2004)

[4] Kit, C., Liu, X.: Measuring mono-word termhood by rank difference via corpus comparison. Terminology **14**(2) (2008) 204–229

[5] Vivaldi, J., Cabrera-Diego, L.A., Sierra, G., Pozzi, M.: Using wikipedia to validate the terminology found in a corpus of basic textbooks. In Calzolari, N., Choukri, K., Maegaard, B., Mariani, J., Odjik, J., Piperidis, S., Tapias, D., eds.: PROC of the 8th INT CNF on Language Resources and Evaluation (LREC), Istanbul, Turkey, ELRA (2012) 3820–3827

[6] Conrado, M.S., Di Felippo, A., Pardo, T.S., Rezende, S.O.: A survey of automatic term extraction for Brazilian Portuguese. Journal of the Brazilian Computer Society (JBCS) **20**(1) (2014) 12

[7] Ahmad, K., Gillam, L., Tostevin, L.: University of surrey participation in TREC8: Weirdness indexing for logical document extrapolation and retrieval (WILDER). In: PROC of the Text REtrieval CNF (TREC). (1999) 1–8

[8] Luhn, H.P.: The automatic creation of literature abstracts. IBM Journal os Research and Development **2**(2) (1958) 159–165

[9] Nogueira, B.M.: Avaliação de métodos não-supervisionados de seleção de atributos para Mineração de Textos. Master's thesis, Institute of Mathematics and Computer Science (ICMC) – University of Sao Paulo (USP), São Carlos, SP (2009)

[10] Salton, G., Yang, C.S., Yu, C.T.: A theory of term importance in automatic text analysis. Journal of the American Association Science **1**(26) (1975) 33–44

[11] Maynard, D., Ananiadou, S.: Identifying terms by their family and friends. In: PROC of 18th INT CNF on Computational Linguistics (COLING), Saarbrucken, Germany (2000) 530–536

[12] Pantel, P., Lin, D.: A statistical corpus-based term extractor. In: PROC of the 14th Biennial CNF of the Canadian Society on COMP Studies of Intelligence (AI), London, UK, Springer-Verlag (2001) 36–46

[13] Zavaglia, C., Aluísio, S.M., Nunes, M.G.V., Oliveira, L.H.M.: Estrutura ontológica e unidades lexicais: uma aplicação computacional no domínio da ecologia. PROC of the 5th Wkp em Tecnologia da Informação e da Linguagem Humana (TIL) – Anais do XXVII Congresso da Sociedade Brasileira da Computação (SBC), Rio de Janeiro (2007) 1575–1584

[14] Lopes, L., Vieira, R.: Aplicando pontos de corte para listas de termos extraídos. In: PROC of the 9th Brazilian Symposium in Information and Human Language Technology (STIL), Fortaleza, Brasil, Sociedade Brasileira de Computação (2013) 79–87

[15] Vivaldi, J., Rodríguez, H.: Evaluation of terms and term extraction systems: A practical approach. Terminology: INT Journal of Theoretical and Applied Issues in Specialized Communication **13**(2) (2007) 225–248

[16] Knoth, P., Schmidt, M., Smrz, P., Zdráhal, Z.: Towards a framework for comparing automatic term recognition methods. In: CNF Znalosti. (2009)

[17] Zhang, Z., Iria, J., Brewster, C., Ciravegna, F.: A comparative evaluation of term recognition algorithms. In Calzolari et al., ed.: PROC of the 6th INT CNF on Language Resources and Evaluation (LREC), Marrakech, Morocco, ELRA (2008) 2108–2113

[18] Souza, J.W.C., Felippo, A.D.: Um exercício em linguística de corpus no âmbito do projeto TermiNet. Technical Report NILC-TR-10-08, Institute of Mathematics and Computer Science (ICMC) – University of Sao Paulo (USP), Sao Carlos, SP (2010)

[19] Coleti, J.S., Mattos, D.F., Genoves Junior, L.C., Candido Junior, A., Di Felippo, A., Almeida, G.M.B., Aluísio, S.M., Oliveira Junior, O.N.: Compilação de corpus em Língua Portuguesa na área de nanociência/nanotecnologia: Problemas e soluções. In of Sao Paulo (USP), U., ed.: Avanços da linguistica de Corpus no Brasil. Volume 1. 192 edn. Stella E. O.Tagnin; Oto Araújo Vale. (Org.), Sao Paulo, SP (2008) 167–191

[20] Bick, E.: The Parsing System "Palavras". Automatic Grammatical Analysis of Portuguese in a Constraint Grammar Framework. University of Arhus, Arhus (2000)
[21] Church, K.W.: One term or two? In: PROC of the 18th Annual INT CNF on Research and Development in Information Retrieval (SIGIR), New York, NY, USA, ACM (1995) 310–318
[22] Salton, G., Buckley, C.: Term weighting approaches in automatic text retrieval. Technical report, Cornell University, Ithaca, NY, USA (1987)
[23] Liu, L., Kang, J., Yu, J., Wang, Z.: A comparative study on unsupervised feature selection methods for text clustering. In: PROC of IEEE INT CNF on Natural Language Processing and Knowledge Engineering (NLP-KE). (2005) 597–601
[24] Dhillon, I., Kogan, J., Nicholas, C.: Feature selection and document clustering. In Berry, M.W., ed.: Survey of Text Mining. Springer (2003) 73–100
[25] Liu, T., Liu, S., Chen, Z.: An evaluation on feature selection for text clustering. In: PROC of the 10th INT CNF on Machine Learning (ICML), San Francisco, CA, Morgan Kaufmann (2003) 488–495
[26] Conrado, M.S., Pardo, T.A.S., Rezende, S.O.: Exploration of a rich feature set for automatic term extraction. In Castro, F., Gelbukh, A., González, M., eds.: Advances in Artificial Intelligence and its Applications (MICAI). LNCS. Springer (2013) 342–354
[27] Estopa, R., Martí, J., Burgos, D., Fernández, S., Jara, C., Monserrat, S., Montané, A., noz, P.M., Quispe, W., Rivadeneira, M., Rojas, E., Sabater, M., Salazar, H., Samara, A., Santis, R., Seghezzi, N., Souto., M.: La identificación de unidades terminológicas en contexto: de la teoría a la práctica. In: Terminología y Derecho: Complejidad de la Comunicación Multilingue, Cabré, T. and Bach, C. and Martí, J. (2005) 1–21
[28] Park, Y., Patwardhan, S., Visweswariah, K., Gates, S.C.: An empirical analysis of word error rate and keyword error rate. In: 9th Annual CNF of the INT Speech Communication Association (INTERSPEECH), INT Speech Communication Association (ISCA) (2008) 2070–2073
[29] Barrón-Cedeño, A., Sierra, G., Drouin, P., Ananiadou, S.: An improved automatic term recognition method for spanish. In: PROC of the 10th INT CNF on COMP Linguistics and Intelligent Text Processing (CICLing), Berlin, Heidelberg, Springer-Verlag (2009) 125–136

Elena Yagunova, Lidia Pivovarova, and Svetlana Volskaya

News Text Segmentation in Human Perception

Abstract: The paper presents a methodology for investigation of human text perception in the process of news comprehension. We conduct four psycholinguistic and two computational experiments to obtain various text segmentations. We consider text elements of various language levels: from words and collocations to syntagmas, from syntagmas to propositions and from propositions to units larger than one sentence. On every language level it is possible to determine operative units, crucial for human text comprehension. We discover these units and then analyze their dependencies to each other and to various types of context.

1 Introduction

We present a methodology for an experimental study of the structural variety of Media texts and the influence of the text structure on text perception and comprehension by humans. In this paper we focus mostly on coherence and cohesion on various language levels, i.e. phenomena that determine text predictability.

Media texts are stylistically heterogeneous; news message, which is our main object, can be described as a subtype of the more general Media text genre. The text structure of a news message is quite similar to the structure of news stream text and defined mostly by traditional written communication. At the same time, the addressee of the news texts is a general audience that makes news so specific.

In our study we consider all language levels: from words and collocations to syntagmas, from syntagmas to propositions, and from propositions to units larger than sentence. Considering all these types of text elements is necessary to obtain a general idea of a text and to study an interaction between units of different levels. The occurrence of some particular units in a text has a probabilistic nature and depends on genre and context properties. Each language phenomenon exists in various context types simultaneously: from the immediate word context to the text and the discourse. All these types of context play their own roles in human perception.

Elena Yagunova, Svetlana Volskaya: Saint Petersburg State University, Saint Petersburg, Russia
Lidia Pivovarova: University of Helsinki, Helsinki, Finland

A word or a collocation is a basic unit for text segmentation. Henceforth we assume that *collocation* is a nonrandom combination of two or more lexical units, which is typical for an analyzed corpus. These typical units can be extracted from the corpus by means of statistical measures, or by experiments with informants.

We define *syntagma* as a unit of perception and comprehension for both written and oral texts. Syntagmatic segmentation, on the one hand, allows communicatee to divide the text into parts which are easy to perceive and, on the other hand, define semantic accents, i.e. the most important elements of the text [9, 10]. Syntagmatic segmentation is ambiguous: it is determined by goals, backgrounds and expectations of the communicatee.

A person uses rather large elements during the process of text comprehension: not only words but also syntagmas and even phrases. It is widely known that increase in size of such elements can cause increase in speed of perception and comprehension processes. However, large operative units can be used by communicatee only if the text is rather predictable; it demands an appropriate knowledge base and skills.

News text, which is the subject of this paper, is quite predictable and easy for human comprehension. We conduct a series of psycholinguistic experiments to find out which operative units are used by a communicatee to adopt the text structure and meaning. We compare these units with automatically extracted structural elements of various sizes. The two sets of results do not always correspond; they are rather two different types of evidence that can be used to study text coherence and cohesion.

As a result of our experiments we cut the text into structural units of various sizes; measure the relative weights of structural units in accordance with various semantic accents (i.e., keywords within the structural unit); analyze the interaction between these structural units and a context, where the context varies from words to text and corpus.

2 Related Works

Identification of discourse structure is crucial problem for automatic text summarization and might be useful for many natural language processing tasks. We focus on linear (not hierarchical) segmentation: the text segments do not overlap [8, 11, 20]. This kind of segmentation reveals the prosodic boundaries with different weights and thus can be useful for speech synthesis task [19, 16].

There are many approaches to automatic detection of text (discourse) structure and its segmentation [8, 13, 14]. The TextTiling algorithm [6, 7] attempts to recognize the subtopic changes using patterns of lexical co-occurrence and distribution. LSITiling [15] algorithm is an extension of the Text-Tiling, where subtopic boundaries are assumed to occur at the points of significant vocabulary shifts. More recently, the TopicTiling algorithm was introduced [17]: this method combines TextTiling with Topic Modelling (LDA) technique.

There are techniques which use clustering and/or similarity matrices based on word co-occurrences [20, 4]. Others use machine learning techniques to detect keywords, or hand-selected keywords to detect segment boundaries [12, 3].

In all the above-mentioned works human judgments are used to evaluate various algorithmic solutions. However, it is not fully clear how humans understand the text and what kind of segments is the most important for this cognitive process. In this paper we are trying to find common tendencies in human text segmentation and perception.

3 Data

We use a highly homogeneous corpus of Russian news devoted to a visit of Arnold Schwarzenegger to Moscow in October 2010; the corpus contains 360 documents, 110 thousand tokens. The event was relatively popular in Russia; we can expect that almost all informants participated in our experiments had at least a little knowledge about this event. The corpus has been obtained automatically using the Galaktika-Zoom News Aggregator[1], which categorizes news articles using text clustering techniques. All the documents are extracted from a daily news stream; it guarantees genre coherence of the corpus. Furthermore, each cluster consists of articles written approximately at the same time, and it ensures a narrative homogeneity as well. The new aggregator produces two types of response: the cluster of news and an “information portrait” [1, 2], which is a set of keywords that distinguish the cluster from the general news stream. The keywords are extracted automatically based on entropy of word distributions in the news stream and a cluster. We use this set of keywords for comparison in Section 5.1.

The choice of this particular cluster as our research corpus is determined by the following reasons: the corpus should have as simple and clear syntactic and semantic structure as possible; it should have a narrative structure, i.e. main character(s), main action, time, place, etc.; it should have a relatively large size to be reliable during computational experiments.

1 The demo version can be found at http://webground.su/

4 Methodology

4.1 Experiments with Informants

We conduct four experiments with informants; each experiment estimates a connectivity level of structural elements of a particular language level. All informants are students, native speakers of the Russian language; each informant participates only in one experiment. The students attended the courses "Introduction to linguistics", "Theory of speech communication" and almost no other linguistic courses.

The *first* experiment is aimed at the extraction of keywords from the text. All the informants had to read the same text from the corpus and write down 10–15 the most important words which describe the text meaning. The group of informants consists of 25 students.

In the *second* experiment informants were asked to scale degree of connectivity between sentences in the text from "0" to "5", where "0" corresponds to the minimal connectivity and "5" – the maximal one. The group of informants consists of 34 students.

In the *third* experiment all the informants had to read the text and to mark syntagmas as understandable and connected "portions" of the text, which have three connection parameters: semantic, syntactic and prosodic. As a result we obtain a syntagmatic structure of the text; key elements of the text meaning are also highlighted. A group of informants consists of 20 students

In the *fourth* experiment informants had to scale the degree of connectivity between *all* words in the text or between word and punctuation symbol from "0" to "5", where "0" corresponds to the minimal connectivity and "5" – the maximal one. The group of informants consists of 21 students.

We had explained to the informants, that we did not examine the "correctness" of the text segmentation or comprehension. The results of these experiments do not always agree with the linguistic tradition, but they are quite informative in the study of subconscious text comprehension by naïve native speakers.

4.2 Computational Experiments

The goal of the *first* computational experiment is to extract keywords from the texts. We use TF-iDF [19], which is the traditional statistic measure, used to estimate the importance of a word in the text which is a part of a larger text collection.

$$TF - tDF(x) = f(x) * \log_2 \frac{1}{f_{d(x)}}, \quad (1)$$

where f(x) is a frequency of x within a given document and $f_{d(x)}$ is a fraction of documents in the corpus that contain x. The largest weights TF-iDF assigned to words, which are frequent within the particular document and but relatively rare in the corpus.

In the *second* computational experiment we investigate the coherence and segmentation of the corpus documents using the open-source Cosegment tool [5]. First, the program collects all the bigrams from a corpus and calculates for them the strength of connectivity using the modified Dice score:

$$Dice'(x, y) = \log_2 \left(\frac{2 * f(x, y)}{f(x) + fg(y)} \right), \quad (2)$$

where f(x) and f(y) are frequencies of the occurrence of x and y in a cluster of texts, and f(x,y) is the frequency of co-occurrence of x and y within the same cluster. The Dice score measures strength of collocates attachment: the score would be maximal if two words never appear in corpus separately.

In the second step, the program scans each document in the corpus and finds in the text connected word chains based on pairwise Dice measure and the immediate context. Finally, the program collects these segments and organizes them according to their frequency in the corpus. Thus, Cosegment produces two outputs: the corpus of texts divided into highly connected segments and the frequency dictionary for these segments.

5 Results

5.1 Experiment 1: Keyword Extraction

As it was mentioned before, we compare three types of keywords: keywords provided by the news aggregation system ("information portrait"), keywords selected by informants, and keywords extracted using TF*iDF measure. The comparison of these sets is presented in Table 1.

As can be seen from the table, the keywords extracted by the majority of informants (the top of the list) correspond to the "information portrait" of the plot (these keywords are highlighted in the table with bold font). The keywords in the middle of the list have more variety: these are both words included into "information portrait" and words, which reflect the sense of a single text (not the whole plot) and the subjective features of comprehension.

Tab. 1: Keywords extracted by informants in comparison with keywords obtained using tf*idf measure; **freq** means the number of informants who selected the word; the **bold font** is used to mark the words that appear in the "information portrait" for the cluster; the *italic* is used to mark the words that appear in the "information portrait' in the different part of speech; the underline is used to mark the words that appear in the first composition fragment of the text.

Informants	freq	tf*idf
Шварценеггер (Schwarzenegger)	23	потребовать (to demand)
Медведев (Medvedev)	21	три-пять (three-five)
технологический (technological)	20	рубль (rouble)
Сколково (Skolkovo)	20	Вексельберг (Vekselberg)
долина (Valley)	13	Russia
Кремниевая (Silicon)	13	вдвоем (two together)
бум (boom)	12	*Кремниевый (Silicon)*
ученые (scientists)	12	установка (aim)
инновационными (by innovation)	11	целевой (goal)
прорыв (breakthrough)	10	миллиард (billion)
разработки (products)	10	объем (volume)
губернатор (governor)	9	прогноз (forecast)
Арнольд (Arnold)	7	возможно (perhaps)
российские (Russian)	7	половина (half)
Дмитрием (by Dmitry)	7	частный (private)
Калифорния (California)	7	автомобиль (car)
американских (of American)	7	бюджетный (budget)
встреча (meeting)	7	средство (mean)
инновационный (innovative)	6	течение (stream)
Президентом (by President)	6	выехать (to leave)
Чайка (Chayka)	5	общий (common)
я (I)	5	*президентский (presidential)*
России (Russia)	5	medvedev@kremlinrussia_e
инвестиционных (investment)	5	направляться (to be going to)
чудо (miracle)	4	салон (cabin)
считает (consider)	4	press
коллеги (colleagues)	4	вчера (yesterday)
ученые (scientists)	4	объявить (to announce)
разработки (products)	4	стремиться (to seek)
аналог (analogue)	4	запечатлеть (to imprint)
компаний (of companies)	4	

Both sets of keywords, which are made during computational experiments – "information portrait" and TF*iDF keywords – pick out a particular unit of analysis in some context. "Information portrait" characterizes a plot (small corpus) in a context of a bigger *news stream*, while TF*iDF extracts words, which distinguish this particular document in a *plot context*[2]. As can be seen from the table, there is almost no overlap between keywords extracted using TF*iDF and either "information portrait" or manually extracted keywords.

A structure of the analyzed plot does not allow TF-iDF to extract words which are important for human comprehension because these meaningful words do not distinguish the text in the context of the plot. This is a significant fact for describing communicatee adaptation for the text peculiarities; it supports the idea that native speakers use broad context – similar to news stream – to comprehend a particular news message.

5.2 Experiment 2: Discourse Degmentation

In this experiment informants estimate the degree of connectivity between sentences in the text. For each pair of the sentences informants grade their connectivity on the scale from 0 to 5; we then use a *median* to estimate final connectivity of the pair. As a result, the text is clearly segmented into three major parts: preamble and start of the plot, where Schwarzenegger comes to Moscow; the plot development, describing various micro-events in which Medvedev and Schwarzenegger participated; and the coda, where they left to the president's residence.

The majority of keywords (17 of 23) appear in the first narrative component that corresponds to traditional news text structure, which assigns the highest weight to the beginning of the text. The keywords that appear in the first fragment are underlined in Table 1. Some sentences in the second, plot development part probably do not have any informational importance for the readers: these are sentences without keywords or with the only keyword.

5.3 Experiment 3: Syntagmatic Segmentation

Let us consider the example of the result of written text segmentation for syntagmas. In the following fragment bold font is used to highlight the keywords extracted by informants. The segment borders are marked with "/". The segments that do not contain any keywords are crossed out. While translating Russian text into English we tried to preserve the segments order as far as it was possible, thus

2 It is possible to use TF-iDF to analyze a larger context: in this case a larger reference corpus should be used to count iDF. But this is beyond our scope in this paper: TD-iDF is used to study a plot context only since we already have an "information portrait" for a larger context.

sentences in the English version of the text do not always follow the normal English word order.

Губернатор *штата* ***Калифорния*** */* ***Арнольд Шварценеггер*** *считает, / что* ***российские ученые*** */ при поддержке* ***американских*** *коллег / смогут совершить в* ***инновационном*** *центре "****Сколково****"* ***технологический прорыв****. / Об этом* ***Шварценеггер*** *заявил во время встречи с* ***президентом России Дмитрием Медведевым****. /* ***Шварценеггер*** *и* ***Медведев*** *встречались летом 2010 года, / когда российский* ***президент*** *посещал* ***Кремниевую долину****. / "****Я*** *тогда вам сказал: / "****Я*** *вернусь". / Вот* ***я*** *и вернулся", / – сказал* ***губернатор Калифорнии****. / "Мне было очень приятно узнать о вашей идее создать аналог* ***Кремниевой долины*** *в* ***Сколково****.* ~~*/ Мы сейчас туда поедем,*~~ */ встретимся с главами* ***американских инвестиционных*** *компаний, / с их* ***российскими*** *партнерами. /* ***Я*** *убежден, / что* ***российские ученые****, / которые занимаются* ***инновационными*** *разработками, / при поддержке* ***американских*** *коллег смогут совершить чудо, / создать настоящий* ***технологический бум****", / – отметил* ***Шварценеггер****. /* ~~*В свою очередь*~~ */* ***Медведев*** *поздравил* ***Шварценеггера*** *с тем, что "****Калифорния****, по сути, вышла из кризиса бюджета". / "****Я*** *считаю, что это ваша победа", – / полагает* ***президент России****. /* ~~*По его словам, / в настоящее время в Москве также происходят перемены. / "У нас тоже много самых разных событий происходит. / Так получилось, / что вы приехали в момент, / когда в Москве нет мэра", /*~~ *– пояснил* ***Медведев****. / "Если бы вы были гражданином* ***России****, /* ~~*то могли бы поработать у нас", / – добавил глава государства,*~~ */ напомнив, что* ***Шварценеггер*** *в январе 2011 года / покидает пост* ***губернатора Калифорнии****. / Затем* ***Медведев*** *и* ***Шварценеггер*** *сели в автомобиль "****Чайка****" / и выехали из подмосковной резиденции* ***президента*** *в* ***Сколково****. /*

Governor *of* ***California*** *State /* ***Arnold Schwarzenegger*** *considers /* ***Russian scientists*** */ with the support of the* ***American*** *colleagues / to be able to make a* ***technological breakthrough*** *in the* ***innovation*** *center "****Skolkovo****". /* ***Schwarzenegger*** *said it during a* ***meeting*** *with* ***Russian President****,* ***Dmitry Medvedev****. /* ***Schwarzenegger*** *and* ***Medvedev*** *met in the summer of 2010, / when the Russian* ***president*** *visited* ***Silicon Valley****. / "Then* ***I*** *said to you: / "****I*** *will be back." / And so* ***I*** *am back," / – said the* ***governor*** *of* ***California****. "I was very pleased to hear about your idea to create an equivalent of* ***Silicon Valley*** *in* ***Skolkovo****. /* ~~*Now we will go there,*~~ */ meet with the heads of* ***American investment*** *companies, / with their* ***Russian*** *partners. / I believe / that* ***Russian scientists*** */ which are engaged in* ***innovative*** *products, / backed by* ***American*** *colleagues can work a miracle, / make the* ***technology boom****, "– said* ***Schwarzenegger****. / By-turn /* ***Medvedev*** *congratulated* ***Schwarzenegger*** *with the fact that California, in fact, is out of the budget crisis. / "****I*** *believe that this is your victory," – / noted the* ***president*** *of* ***Rus-***

sia. / ~~*According to him, now in Moscow changes also take place. / "We also have a lot of different events. / So happens / that you have arrived at a time / when Moscow has no the mayor,"*~~ / *– said* ***Medvedev.*** / *"If you were a citizen of* ***Russia,*** / *you could work with us," / – said the head of state, / reminding that* ***Schwarzenegger*** *in January 2011 / is stepping down as* ***governor*** *of* ***California.*** / *Then* ***Medvedev*** *and* ***Schwarzenegger*** *got into the car* ***"Chayka"*** / *and left the* ***presidential*** *residence near Moscow in* ***Skolkovo.*** /

Using keywords we can estimate supposed weight of each syntagma. As can be seen from the fragment above, 6 syntagmas do not contain any keywords, 10 syntagmas contain only one keyword. Weight of syntagma also depends on keyword position in syntagma: initial position or ending point has higher weight in comparison with middle one.

5.4 Experiment 4: Text Coherence and Cohesion

In this experiment informants estimated the degree of connectivity between words in the text. For each pair of the tokens informants grade their connectivity on the scale from 0 to 5; we use a *median* to estimate final connectivity of the pair. We present the same text segmented as a result of this experiment[3]; (___) means the border of connected segment: the connectivity between words within parenthesis is 5 (maximal); ||means the break, the connectivity is 0 or 1; [___] means segmentation obtained in computational experiment with Cosegment program.

([Губернатор штата Калифорния)]([Арнольд Шварценеггер] считает)[, что] ([российские ученые]) ([при поддержке]) ([американских] [коллег) (смогут совершить]) в ([инновационном центре]["Сколково"]) ([технологический прорыв]). ([Об этом]) (Шварценеггер заявил) ([во время] [встречи с президентом] России) ([Дмитрием Медведевым]). Шварценеггер и Медведев [встречались летом] ([2010 года], [когда]) ([российский президент] посещал) [Кремниевую долину.] || "Я [тогда (вам сказал]): (["Я вернусь]). [Вот я и вернулся"], ([– сказал][губернатор Калифорнии]). ||"[Мне было] (очень [приятно) (узнать] о) ([вашей идее]) [создать аналог] ([Кремниевой долины]) ([в Сколково.]) Мы сейчас ([туда поедем]), ([встретимся с] [главами) американских (инвестиционных компаний]), с [их (российскими партнерами]). ([Я убежден,] [что])

3 It would be hard to translate these results into English because the segmentation is highly depends on micro-syntax structure. Hopefully, the visual clues – the length of the segments, the distribution of keywords – may give an idea of the potential of this methodology.

([***российские ученые***]), ([*которые занимаются*) (***инновационными** разработками*]), ([*при поддержке*]) (***американских*** [*коллег*) (*смогут*) [*совершить чудо*]), [*создать настоящий* (***технологический бум***]) [", – *отметил*] ***Шварценеггер***. *В* ([*свою очередь*]) (***Медведев*** [*поздравил* ***Шварценеггера***]) ([*с тем*])[, *что*] "***Калифорния***, ([*по сути*]), [*вышла* (*из*)[*кризиса*) *бюджета*]". ||

(["***Я*** *считаю*][, *что*]) *это* ([*ваша победа*])", – *полагает* ([***президент России***]). [*По его словам*], *в* ([*настоящее время*]) ([*в Москве*]) *также* ([*происходят перемены*]). (["*У нас*]) [*тоже много*] [*самых разных событий происходит.*] ([*Так получилось*])[, *что*] ([*вы приехали*]) ([*в момент,*]) *когда* (*в Москве*) [*нет мэра*]", (– *пояснил* ***Медведев***.) "*Если бы* (*вы были* [*гражданином* ***России***]), [*то* (*могли*][*бы*) *поработать*] ([*у нас*])", – *добавил* ([*глава государства*), *напомнив*[, *что*] ***Шварценеггер*** (*в* [*январе*) (*2011 года*]) [*покидает* (*пост*] [***губернатора Калифорнии***.]) *Затем* [***Медведев*** *и*] ***Шварценеггер*** ([*сели в*] *автомобиль*) ("***Чайка***") (*и* [*выехали из*) (*подмосковной резиденции*]) ***президента*** ([*в* ***Сколково***.])

It can be seen from the fragment that computational segmentation in many cases corresponds to the segmentation obtained in psycholinguistic experiment. However, the computational segments are shorter and in some cases non-grammatical.

The data of this experiment may be combined with the results of the experiments 2 and 3: the more keywords are in the segment, the higher weight of the segment. For example, a segment *during the **meeting** with the **President** of **Russia*** has a high weight due to the fact that a) there are three keywords in the segment, b) the length of the segment is 6 words. Occurrence of the keyword ***President*** in this string also has a higher weight in comparison with, for example, a single occurrence of this word in the string *residence of **President** near Moscow*.

6 Conclusions

In this paper we have presented a methodology that combines psycholinguistic and computational experiments. We applied this methodology to the small homogeneous corpus of news texts, and to one particular document within the corpus. As a result, the document is segmented into structural units of various scales and annotated with operative units crucial for human text comprehension.

A set of keywords, which is selected during experiment with informants, represents text folding and shows the peculiarities of news texts perception and comprehension by naïve speaker. Quite similar set of keywords can be extracted automatically but this task requires a larger text collection as an input.

The hierarchy of keywords and their distribution in the text reflect semantic structure of the text: the most important keywords occur in the opening fragment of the text. Comparison of two keywords sets – extracted automatically and selected by informants – characterizes the specificity of the analyzed text in a context of the plot and reflects the peculiarities of perception of the text.

Keywords (e.g. nominations) in the news texts are often emphasized by syntagma bounds, and sometimes syntagma bounds also form a border between topic and focus components. A proposition generally coincides with sentence in the news texts, though it could be less than a sentence (e.g., clause).

Propositions join together and form discourse units. This language level allows us to describe and classify a text as, for example, a simple event or a sequence of events bound by a cause-effect relation.

The results obtained by this methodology can be further verified by experiments, where informants have to restore the text content using the segments of various length and weight. The methodology might be useful i) to compare human perception of various genres and topics ii) to study variety among people in sociolinguistic studies iii) to obtain a "gold standard" for automatic text segmentation; iv) to annotate corpora with cognitive units. However, in two latter cases it would be necessary to implement a lighter version of the method, which would involve fewer informants.

Acknowledgement

The authors acknowledge Saint-Petersburg State University for a research grant 30.38.305.2014.

Bibliography

[1] Antonov, A.V., Bagley, S.G., Meshkov, V.S., Sukhanov, A.V. (2006) Clustering of the documents on the base of metadata. *Proceedings of the international conference Dialogue'2006*, Moscow, Russia, 31 May – 4 June.

[2] Antonov A.V., Yagunova E.V. Procedure of working with text information collections via information portraits analysis / *Proceedings of RCDL'2010*, pp. 79–84 [In Russian].

[3] Beeferman D., Berger A. and Lafferty J. (1997) Text Segmentation Using Exponential Models. *Proceedings of the Second Conference on Empirical Methods in Natural Language Processing*, Association for Computational Linguistics, Brown University, August 1997.

[4] Choi, F. Y. Y. (2000) Advances in domain independent linear text segmentation. *Proceedings of the North American Chapter of the Association of Computational Linguistic*, Seattle, Wash., 29 April – 3 May 2000.

[5] Daudaravicius, V. (2010) Automatic identification of lexical units. *Computational Linguistics and Intelligent text processing CICling-2009*, Meksikas, Meksika.

[6] Hearst M. A. (1994) Multi-Paragraph Segmentation of Expository Text. *Proceedings of the 32nd Annual Meeting of the Association for Computational Linguistics,* Las Cruces, New Mexico, 27–30 June 1994.

[7] Hearst M. A. (1997) TextTiling: Segmenting Text into Multi-paragraph Subtopic Passages, *Computational Linguistics*, 23(1): 33–64.

[8] Kan, Min-Yen, Klavans, J. L., McKeown, K. R. (1998) Linear segmentation and segment significance. *Proceedings of WVLC-6*, Montreal, Canada, August 1998.

[9] Krivnova, O.F., Chardin, I.S. (2002) *Pausing in natural and synthetic speech.* Donets.

[10] Krivnova, O.F. (2007) *Rhythmization and prosodic segmentation of text during the process "speech-thought" (theoretical and experimental research)*, PhD hab. Thesis, Moscow.

[11] Marcu D. (1997) The Rhetorical Parsing of Natural Language Texts. *Proceedings of the 35th Annual Meeting of the Association for Computational Linguistics*, Madrid, Spain, 7 – 12 July 1997.

[12] Passonneau R. J. and Litman D. J. (1993) Intention-based segmentation: human reliability and correlation with linguistic cues. *Proceeding of the 31st Annual Meeting of the Association of Computation Linguistics*, Ohio, USA, 22–26 June 1993.

[13] Pevzner L. and Hearst M. (2002) A critique and improvement of an evaluation metric for text segmentation. *Computational Linguistics*, 28(1):19–36.

[14] Peshkov, K., Prévot, L. Segmentation evaluation metrics, a comparison grounded on prosodic and discourse units. *Proceedings of 9th Conference on Language Resources and Evaluation*, Reykjavik, Iceland, May, 2014.

[15] Řehůřek, R. (2007) Text segmentation using context overlap. *Proceedings of the aritficial intelligence 13th Portuguese conference on Progress in artificial intelligence*, Guimarães, Portugal, December 03–07, 2007.

[16] Reichel, Uwe D., Pfitzinger, Hartmut R. (2006) Text Preprocessing for Speech Synthesis. *Proceedings of the TC-STAR speech to speech translation workshop*. Barcelona, Spain, June 19–21, 2006.

[17] Riedl, M. & Biemann, C. (2012) TopicTiling: A Text Segmentation Algorithm Based on LDA. *Proceedings of ACL 2012 Student Research Workshop*, Jeju Island, Korea, July 09–11, 2012.

[18] Salton G., Buckley C. (1988) Term-weighting approaches in automatic text retrieval. *Information Processing and Management*, 24(5): 513–523.

[19] Taylor, P. (2009) *Text-to-Speech Synthesis*. Cambridge University Press, UK.

[20] Yaari Y. (1997) Segmentation of Expository Text by Hierarchical Agglomerative Clustering. *Recent Advances in NLP 1997*, Tzigov Chark, Bulgaria, September 1997.

Mai Yanagimura, Shingo Kuroiwa, Yasuo Horiuchi, Sachiyo Muranishi, and Daisuke Furukawa

Preliminary Study of TV Caption Presentation Method for Aphasia Sufferers and Supporting System to Summarize TV captions

Abstract: We conducted an experiment in which people with aphasia watched TV news with summarized TV captions made by speech-language-hearing therapists in accordance with our rules for summarizing captions. As a result, we found the summarized TV captions are better subjectively for aphasia sufferers. To familiarize people with the summarized captions, we've developed a supporting system for any volunteers to summarize Japanese captions while keeping a consistent summarization style. A result of preliminary evaluation shows some of symbols applied automatically in initial summarized sentences are effective for summarization manually. Furthermore, we found 40% isn't suitable as a length of initial sentence.

1 Introduction

Aphasia is a medical condition that affects an individual's ability to process language, including both producing and recognizing language, as a result of a stroke or head injury. Though they do not have reading difficulties related to vision or hearing difficulties related to auditory sense, aphasia sufferers encounter several difficulties in understanding written text and speech.

To support their comprehension, the people supporting them like speech-language-hearing therapists (ST) and their family must know about their daily knowledge of communication. They can combine short sentences and words with symbols and pictures to enhance understanding because it is easier for aphasia

Mai Yanagimura, Shingo Kuroiwa, Yasuo Horiuchi: Chiba University, Graduate School of Advanced Integration Science, 1-33, Yayoicho, Inage-ku, Chiba-shi, Chiba, 263-8522 Japan, {mai}@chiba-u.jp
Sachiyo Muranishi, Daisuke Furukawa: Kimitsu Chuo Hospital, 10-10 Sakurai, Kisarazu-shi, Chiba, 292-8535 Japan

sufferers with reading difficulties to understand writing in Kanji which convey a meaning rather than in Kana, syllabic writing and short and simple sentences rather than long and complex sentences. (Kane et al., 2013) explored ways to help adults with aphasia vote and learn about political issues using common knowledge. (Kobayashi 2010) reports volunteer activities, summary writing, and point writing of speeches in conferences consisting mainly of aphasia sufferers. These activities also use the knowledge of communication. Volunteers type summarization sentences on a projection screen during the conference on summary writing. Next they give overviews by point writing. They prepare summarization sentences in advance before the conference and write new information on the projection and tablet screens during the conference. One of the problems in summary writing is the shortage of STs and volunteers relative to aphasia sufferers. The other is that summary writing is difficult for people other than STs and volunteers because there are no specific guidelines for it.

A summarization technique needs to be applied in daily life, specifically summarization of TV captions. Therefore, we aim to develop an automatic caption summarization system. (Napoles et al., 2011; Cohn et al., 2009; Knight et al., 2002; Boudin et al., 2013) have researched automatic sentence compression methods in English and other languages. However, a practical automatic caption summarization is difficult to make in Japanese. (Ikeda et al., 2005) found readability to be 36% and meaning identification to be 45% for sentences they used. Though (Norimoto et al., 2012) showed the 82% of sentences they used could be correctly summarized, the remaining 18% of sentences has their meanings changed and did not work as sentences. These pieces of research were for deaf and foreign exchange students. (Carrol et al., 1998) developed a sentence simplification system in English that was divided into an analyzer and a simplifier for people with aphasia. However, we cannot find a sentence compression system for aphasia sufferers in Japanese.

Therefore, before developing an automatic summarization system, our purpose here is to develop a supporting system to help people summarize TV captions manually. Fig. 1 shows the concept of our system to provide videos with summarized captions. We build a server to manage original and summarized captions, videos, and editing logs. We obtain videos and original captions from video providers. We provide the videos and captions to supporting people who use our supporting system to summarize captions. They summarize initial summarized text that has been generated automatically in our system. We can then provide videos with summarized captions for people with aphasia. Although (Pražák et al., 2012; Imai et al., 2010; Hong et al., 2010) proposed captioning system using speech recognition for deaf, it is hard to find research work for people with aphasia such our developing system.

We also use sentence compression to generate the initial summarized text. Ex. 1 presents an original sentence and its automatically and manually summarized version.

[Ex. 1]
Original Caption:
今日は有栖川宮親王がドライブで訪れたという 国立市の神社でレプリカのお披露目が行われました
(A replica car was unveiled today at a shrine Prince Arisugawa-no-miya once he visited in Kunitachi city)

Summarized TV caption:
《国立市　神社》レプリカ お披露目

(Replica unveiled 《Shrine in Kunitachi》)

We need to paraphrase from Kana into Kanji aggressively and use symbols to replace deleted words and phrases in such sentences. Previous research did not summarize these features. Therefore, we need additional methods to generate initial summarized text for people with aphasia.

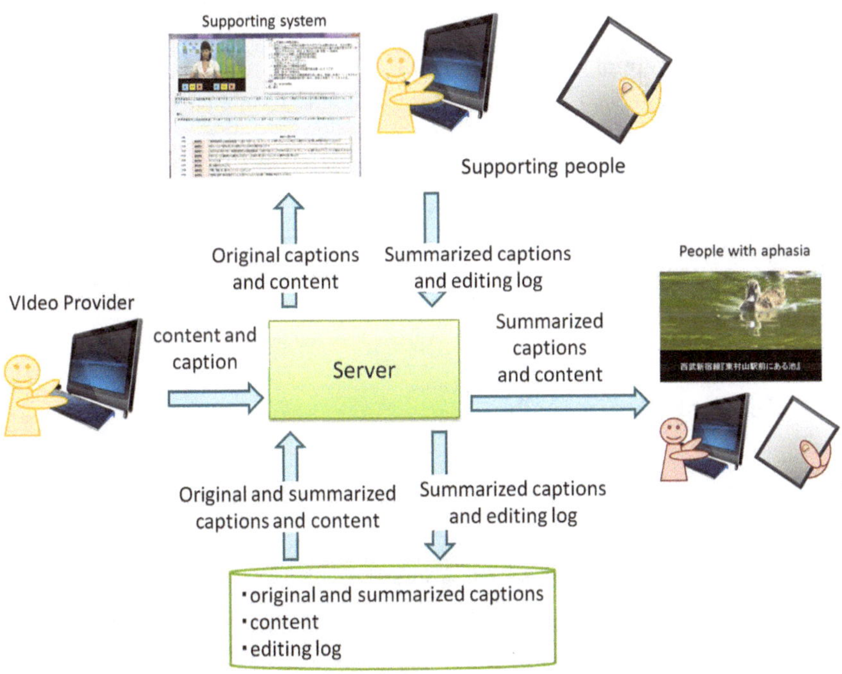

Fig. 1: Concept of our project.

After that, we evaluate the effect of summarized TV captions because we could not find a comprehensive study that discusses the effects of summarized TV captions for aphasia sufferers in Japanese. In Sec.3, we perform an experimental evaluation of summarized TV captions for aphasia sufferers to compare them with original captions in Sec.3. Next, in Sec.4, we detail a method to generate the initial summarized captions for people with aphasia. Finally, we evaluate the system in Sec. 5 and make some concluding remarks Sec.6.

2 Guidelines of Summarized TV Captions for Aphasia Sufferers

We have proposed the following guidelines for manually producing accessible summarized captions for adults with aphasia using the knowledge of summary writing and communication of STs. These guidelines consist of 27 rules.

2.1 Overview of Guideline

Summarized TV captions have the following features.
1. Shorten summarized TV captions to 50–70% of original captions.
2. Summarize as clearly and concisely as possible
3. Paraphrase Kana into Kanji
4. Use some symbols to replace words and phrases

We define various specific rules: paraphrasing; deletion and usage of symbols to summarize sentences. In this paper, we present a part of all rules as follows.

2.2 Rules of Paraphrasing

These rules aim to transform writing in Kana into Kanji.

(1) Into smaller chunks including more Kanji

[Ex.]

Original caption: めちゃくちゃに壊れていました.

(It was collided violently.)

Summarized caption: 大破

(Crash.)

(2) Into nouns from general verbs
[Ex.]
Original caption: 乗用車が見つかりました.
(They have discovered a car unexpectedly.)
Summarized caption:『乗用車』発見
(『Car』found.)
(3) Into general expressions from honorific ones
[Ex.]
Original caption: 佐藤さんがいらっしゃいました.
(Mr. Sato has arrived.)
Summarized caption: 佐藤　来た
(Sato came.)
(4) Into ending a sentence with a noun or noun phrase
[Ex.]
Original caption: 地元の人達の支援を受けてすくすくと成長しています.
(They are growing up healthily with the support of local people.)
Summarized caption: 地元の人達 支援 → 成長中
(Growing up ← local people's support.)

2.3 Rules of Deletion

The rules aim to decrease writing in Kana.
(1) Delete fillers
[Ex.]
Original caption: えーと無事に 9 羽あの一飛び立ってほしいということですよね.
(Uh, we wish that, well, all 9 birds fly away safely.)
Summarized caption:「無事に 9 羽飛び立って！！」
(「Nine birds fly away safely!」)
(2) Delete postpositional particles
[Ex.]
Original caption: 主人公の犬を演じるのは高校 2 年生熊田里菜さんです.
(The leading role of the dog is to be played by a second-year high school student, Ms. Kumada Rina.)
Summarized caption: 主人公の犬 ＝ 高校 2 年生『熊田里菜』
(Lead role of dog = second-year HS student『Kumada Rina』)

(3) Delete terms you decide can be inferred
[Ex.]
Original Caption: 皆様の多大なご声援により，埼玉県で初めて今回のフードイベントが行われました．
(This food event was held in Saitama Prefecture for the first time thanks to your enthusiastic encouragement.)
Summarized caption:《埼玉県》フードイベント初開催
(《Saitama Prefecture》 hosts first food event .)
(4) Delete terms you decide are not relevant to key content
[Ex.]
Original Caption: カルガモへのメッセージを記してもらうノートを置き、駅を利用する市民などが書き込んでいました．
(Local people placed a notebook to write messages for baby spot-billed ducks, and station users wrote, “Good luck” and “Grow up healthily”.)
Summarized caption:『カルガモへメッセージノート』設置
(『Massage note for baby ducks』placed)

2.4 Rules of Symbols

We use symbols to paraphrase language and emphatic expressions effectively.
(1) 《 》: Enclose location and time expressions
[Ex.]
Original caption: 西武新宿線の東村山駅前にある池では毎年カルガモのひなが生まれています.
(Baby spot-billed ducks are born annually in a pond in front of Higashimurayama station on the Seibu-Shinjuku Line.)
Summarized caption: 《西武新宿線 東村山駅》《毎年》カルガモのひな 誕生
(《Annuall》 baby ducks are born 《Higashi-murayama station on Seibu-Shinjuku Line》.)
(2) 『 』: Enclose key terms to emphasize them
[Ex.]
Original caption: 誕生して間もないカルガモのひな 9 羽が地元の人達の支援を受けてすくすくと成長しています.
(Nine newborn baby spot-billed ducks are growing up healthily with the support of local people.)
Summarized caption:『カルガモのひな（9 羽）』地元の人達 支援 → 成長中
(『Baby spot-billed ducks (nine)』growing up ← local people’s support.)

(3) ( ) : Enclose less important information
[Ex.]
Original caption: 誕生して間もないカルガモのひな 9 羽が地元の人達の支援を受けてすくすくと成長しています.
(Nine newborn baby spot-billed ducks are growing up healthily with the support of local people.)
Summarized caption:『カルガモのひな（9 羽）』地元の人達 支援 → 成長中
(『Baby spot-billed ducks (nine)』growing up ← local people's support.)

(4) →: Paraphrasing Conjunction and conjunctive particles to express addition
[Ex]
Original caption: 9 羽は今では大人の手のひらを超える大きさに成長し今日も駅前の池で泳ぎ回っています.
(The nine baby birds have been grown up to palm-size and are swimming in a pond in front of the station.)
Summarized caption:『9 羽』大人の手以上の大きさ → 《駅前の池》遊泳
(『Nine birds』grown up to palm-size → swimming 《pond in front of station》.)

(5) ⇒: Paraphrasing → into end of sentence when → is already used in the sentence
[Ex.]
Original caption: 低気圧が北上し関東地方では雪が降りますので足元にお気を付け下さい.
(Please take care because low pressure will go up north, leading to snow in the Kanto region.)
Summarized Caption: 低気圧 北上 → 《関東地方》雪 ⇒ 足元注意
(Low pressure going up north → 《Kanto region》 snow ⇒ Take care.)

3 Summarized TV Caption of Evaluation

We performed an evaluation of summarized TV captions for people with aphasia. Fig. 2 shows the evaluation procedure. We manually summarized TV captions for 4 TV news segments in accordance with the guidelines. We presented two types of content (with the original or summarized TV captions) to 13 aphasia sufferers. The summarized captions were assessed in intelligibility tests, subjective evaluations, and interviews. T1, T2, T3 and T4 represented four TV news segments. The number of sentences in T1~T4 was 36 sentences. Each TV segment was about 1 minute and 30 seconds long. Aphasia sufferers tried the intelligibility test after watching each news with original captions or summarized captions. Each aphasia suffer tried the intelligibility test four times as shown in Fig. 2. They filled out the answer sheet for some questions about the news they watched just before the tests. We asked "Which do you think is better and why: the TV caption or the summarized TV caption?" in the subjective evaluations after watching 2 news with original captions and summarized captions. Each aphasia suffer tried the subjective evaluation twice as shown in Fig. 2.

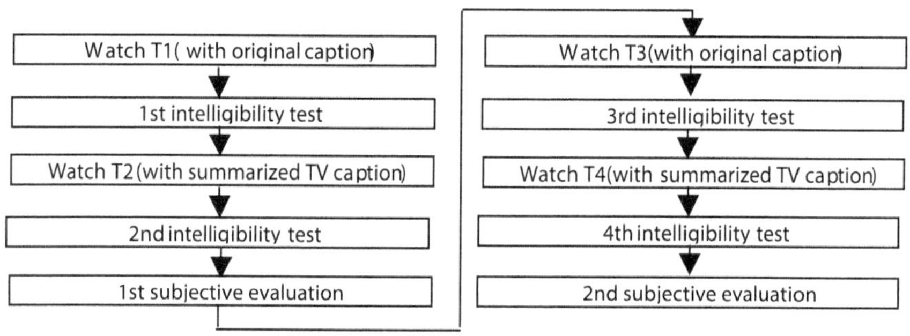

Fig. 2: Example procedure for one person.

We organized some procedures to prevent the order effect. We studied this evaluation in a hospital where the subjects were undergoing rehabilitation. This experiment lasted 30 minutes per person. The contents included topics such as birds, a car accident, and so on.

3.1 Experimental Results and Discussion

Eight out of 13 people the first time and nine out of 13 people the second time replied "The summarized captions are better". We received some positive opinions from the aphasia sufferers, for example, "They are precise" and "They're easy to read". These results suggest that summarized captions are suitable to give an overview of news, not detailed information. Furthermore, although content was about only 1 minute and 30 seconds, some people remarked "I become tired watching content with the original captions" in interviews. This suggests that original captions are not simple enough for people with aphasia to watch contents for a long time. Therefore, we found summarized TV captions can be one effective option for closed caption expressions for aphasia sufferers.

Table 1 presents the average correct percentages in the intelligibility test of people who said the summarized or original TV captions are better. The percentage for summarized TV captions is lower than was expected. (There is no significant difference in T test.) We discuss why the average correct percentages of the intelligibility test are lower for aphasia sufferers who preferred summarized captions. Some were confused by the symbols used to summarize captions because we did not explain their meanings before this experiment. One aphasia sufferer remarked “I don’t know the meanings of the symbols” in the interview. We need to familiarize aphasia sufferers with the meanings of symbols.

Tab. 1: Average percentage of correct answers in intelligibility test.

	People who prefer summarized captions	People who prefer original captions
Original caption	0.68	0.69
Summarized caption	0.65	0.54

Another reason is the vague usages of brackets, 『 』. We proposed the following rule for『 』: you enclose key terms with『 』to emphasize them. Although『乗用車』and 乗用車 are the same key term (*joyosha*: car), 『 』was used for the former but not for the latter. This is why it is not enough for one person only to apply rules consistently. For this reason, our system supports people to summarize captions. Ex. 1 presents a sentence that is difficult to understand because we use too many arrows →. This could hinder reading comprehension. We should consider how to use → and how many to use in more detail in order to generate better sentences like Ex. 2.

[Ex. 1]
大型トラック トレーラー追突 → 炎上→『1人死亡』⇒
(Big Rig collided with tractor-trailer → burning → man killed)

[Ex. 2]
大型トラック トレーラー追突→炎上
(Big Rig collided with tractor-trailer → burning man killed.)

4 Supporting System to Summarize

We found that summarized captions are an option for closed captions in Sec. 3. We thus developed a system to make summarized captions known and overcome the shortage of people to summarize. Fig. 3 shows the two-phase procedure of the system. The system summarizes TV captions in the automatic summarization phrase to generate initial summarization captions. Then, in the manual summa-

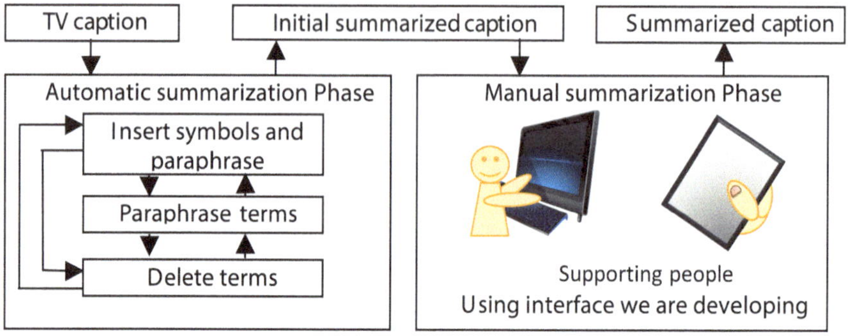

Fig. 3: Procedure of supporting system to summarize captions.

rization phase, volunteers summarize initial summarized captions on the PC or tablet using an interface we are developing.

4.1 Rule for Supporting System

We detail methods to generate initial summarized captions in this section. The system summarizes captions in accordance with the 27 rules in Sec. 2 written in XML. We used Headline style summarization (Ikeda et al. 2005) and a deletion method (Norimoto et al. 2012) to apply paraphrasing and deletion rules like paraphrasing to end a sentence with a noun or noun phrase. Moreover, we determine a new method to apply some symbol rules, 《 》,『 』, →, ⇒ because these symbols account for approximately 80% in all symbols in summarized captions in Sec. 2. We used CaboCha (Kudo et al., 2003) as a Japanese dependency structure analyzer and MeCab (Kudo et al., 2004) as a Part-of-Speech and Morphological Analyzer. Furthermore, we manually made a database of 203 conjunctions and conjunctive particles and tagged them. We describe a part of all methods to insert or paraphrase each symbol in below.

I. **《 》: Enclose location and time expressions**
   If a chunk including phrases tagged LOCATION or TIME by Cabocha.

II. **『 』: Enclose key terms to emphasize them**
   If the tfidf score defined by Norimoto terms is over the threshold, the system encloses the terms. Furthermore, we delete all words except nouns inside the chunk.

III. **→ : Paraphrase conjunction and conjunctive particles to express addition**
   If a word matches conjunctions or conjunctive particles to express addition in the database, the system paraphrases the word into an arrow →.

IV. → : Paraphrase resultative conjunction and conjunctive particles to express causality
If a word matches resultative conjunctions or conjunctive particles to express causality in the database, the system paraphrases the word into an arrow →.

V. ⇒ : Paraphrasing → into end of sentence when → is already used in the sentence
If some → have been applied already in sentence, the system paraphrases the last → into ⇒.

4.2 Preliminary Experiment to Evaluate Supporting System

We performed an experimental evaluation of the supporting system. The evaluation was intended to determine the most suitable summarization percentage for editing. We made between 100–40% summarization that appeared in the edit box, from which user made a summarized caption. In 100% condition, the sentences are applied all symbols and paraphrasing rules before the experiment. One student subject summarized the same original captions in Sec. 3 using the system. The system was assessed by a user-interview about the usefulness of each initial summarized sentence.

Sentences in 100%, 80% and 60% condition were chosen as useful sentences. The student suggested some reasons to choose useful sentences. One of the reasons is that 《 》 had already been applied correctly in initial summarized sentences. We use 《 》 to enclose location and time expressions in our guideline. Ex. 1 shows a useful initial summarized sentence and its original sentence because 《 》 is applied correctly as《群馬県富岡市の上信越自動車道》in initial summarized caption. 群馬県富岡市の上信越自動車道 (meaning: at Joshin-etsu Expressway in Gunma) in original caption means a location expression.

Another reason is that → and ⇒ gave points of beginning summarization. We use → to paraphrase conjunction and conjunctive particles to express addition and ⇒ to paraphrase → into end of sentence when → is already used in the sentence. In [Ex. 1], 追突 → 炎上 → 一人が死亡した事故で ⇒ (meaning: a traffic accident which the truck collided with the trailer and went up in flames and a man dies) in the initial caption is used → and ⇒. The students could instantly find point of beginning summarization around → and ⇒ not be worried seriously.

[Ex. 1]
Original caption: 群馬県富岡市の上信越自動車道の下り線で大型トラックがトレーラーに追突して炎上し一人が死亡した事故でその後２台の間に乗用車が挟まれていたことがわかりました。

(We found a car between a heavy duty truck and a trailer after a traffic accident which the truck collided with the trailer and went up in flames and one fatality was recorded at Joshin-etsu Expressway in Gunma Prefecture.)

Initial summarized caption: 《群馬県富岡市の上信越自動車道》の下り線で『大型トラック』が『トレーラー』に追突 →炎上→一人が死亡した事故で⇒２台の間に乗用車が挟まれていたことが判明
(We found car between heavy duty truck and trailer after traffic accident which 『truck』collided with『trailer』→ burning → a man dies 《Joshin-etsu Expressway in Gunma》.)

Most of sentences chosen as not useful sentences were in 40% condition. The student remarked it was annoying for him to re-insert words deleted already in initial summarized sentences if deleted words were too many. Ex. 2 presents a not useful initial summarized caption in 40% condition and its original sentence. 首に軽いけがをしました (the trailer driver was injured on his neck lightly) in the original caption was deleted in the initial summarized caption. There was too much deletion.

[Ex. 2]
Original caption: 大型トラックの運転席から運転手とみられる１人が遺体で見つかりトレーラーの運転手も首に軽いけがをしました。
(One fatality consider to the heavy truck driver was recorded in the driver's seat and the trailer driver was injured on his neck lightly.)

Initial summarized caption:トラック運転席から運転手と『トレーラー』運転手も
(driver in the seat and『trailer』driver.)

These results show we do not need to use 40% to show in extension experiment we will perform.

5 Conclusion

In this paper, we discussed an evaluation of summarized TV captions and a supporting system to summarize them. First, we experimentally evaluated TV captions summarized for people with aphasia. We, including two speech-language-hearing therapists (STs), proposed summarization guidelines in order to assist summarization of TV captions in Japanese. After that, aphasia sufferers evaluated captions summarized in accordance with the guidelines manually in comparison with original captions. As a result, we found some people with aphasia preferred the summarized captions when they watched the TV contents. Importantly, we should provide the summarized TV captions as one effective option for closed caption expressions for aphasia sufferers.

To familiarize people with the summarized captions, we have developed a supporting system for any volunteers to summarize Japanese TV captions while keeping a consistent summarization style. Our system used some existing sentence compression methods to apply paraphrasing and deletion methods from previous research. What is more, we proposed new methods to apply symbols. The results of the experimental evaluation showed it was effective to summarize captions if ⟨⟨ ⟩⟩, → and ⇒ had already been applied in initial summarized sentences. We also found that we don't need to show sentences whose length is 40% in an extension experiment we will perform soon. We will immediately upload our system on the web and increase the number of volunteers to summarize captions using it.

Acknowledgements

We thank Escor Inc. and Hoso Bunka Foundation for partially funding this research.

Bibliography

Ikeda, S. (2005) Transforming a Sentence End into News Headline Style. In: Yamamoto, K., *Proceedings of the Third International Workshop on Paraphrasing (IWP2005)*, Jeju Island, South Korea, October 14, 2005, pp. 41–48.

Kobayashi, H. (2010) Participation restrictions in aphasia, *Japanese journal of Speech, Language, and Hearing Research*, 7(1):73–80. (In Japanese)

Norimoto, T. (2012) Sentence contraction of Subtitles by Analyzes Dependency Structure and Case Structure for VOD learning system. In: Yigang, L., Koyama, N., Kitagawa, F. (eds.), *Language, IEICE Technical repor.ET*, 110(453):305–310. (In Japanese)

Kane, S.K. (2013) Design Guidelines for Creating Voting Technology for Adults with Aphasia. In Galbraith, C. (ed.), *ITIF Accessible Voting Technology Initiative* #006

Carrol, J. (1998) Practical Simplification of English Newspaper to Assist Aphasic Readers. In: Minnen, G., Canning,Y., Devlin, S. Tait, J. (eds.), *Proceedings of Workshop on Integrating AI and Assistive Technology*, Medison, Isconsin, July 26–27, 31, pp. 7–10.

Hong, R. (2010) Dynamic captioning: video accessibility enhancement for hearing impairment. In: Wang, M., Xu, M., Yan, S., Chua, T. (eds.), *Proceedings of the International Conference on Multimedia*, Firenze, Italy , October 25–29 2010, pp. 421–430.

Pražák , A. (2012) Captioning of Live TV Programs through Speech Recognition and Re-speaking In: Loose. Z., Trmal, J., Psutka, J.V., Psutka, J. (eds.), *Text, Speech and Dialogue Lecture Notes in Computer Science*, 7499:513–519

Imai, T. (2010) Speech recognition with a seamlessly updated language model for real-time closed-captioning In: Homma, S., Kobayashi, A., Oku, T., Sato, S. (eds.), *Proceeding of INTERSPEECH*, Florence, Italy, 15 August 2011, pp. 262–265.

Napoles, C. (2011) Paraphrastic sentence compression with a character-based metric: tightening without deletion In: Burch, C. C., Ganitkevitch, J., Durme, B. V. (eds.), *Proceedings of the Workshop on Monolingual Text-To-Text Generation*, Portland, OR, 24 June 2011, pp. 84–90.

Cohn, T. A. (2009) Sentence Compression as Tree Transduction In: Lapata, M. (ed.), *Artificial Intelligence Research*, 34:637–674.

Knight, K. (2002) Summarization beyond sentence extraction: A probabilistic approach to sentence compression In: Marcu, D. (ed.), *Artificial Intelligence*, 139(1):91–107.

Boudin, F. (2013) Keyphrase Extraction for N-best Reranking in Multi-Sentence Compression In: Morin, E. *Proceedings of North America Chapter of the Association for Computational Linguistics*, Atlanta, USA, June 9th-15 2013, pp. 298–305.

Olga Acosta and César Aguilar

Extraction of Concrete Entities and Part-Whole Relations

Abstract: This paper is focused on extraction of part-whole relations where concrete entities are involved. These concrete entities are identified considering their axial properties. In a spatial scene something is located by virtue of the axial properties associated with a reference object. In this case, reference objects can be concrete entities. In linguistic terms, axial properties are represented by place adverbs. Additionally, for identifying reference objects in a sentence we consider syntactical patterns extracted by means of a chunking phase. In order to remove noise in the results, we take into account most frequent nouns in a general language reference corpus and descriptive adjectives obtained by linguistic heuristics. Finally, concrete entities can be used for locating other entities related by a part-whole relation. Results show a low precision, but a good recall for detecting concrete entities relevant to a specialized domain.

1 Introduction

The identification of terms inserted into lexical-semantic relations is a way explored by several researchers in the recent years. A good example of this exploration is the volume prepared by Auger and Barrière (2010), which join a set of works oriented to discover these kinds of relations in order to extract terms, definitions and other conceptual units.

According to Buitelaar, Cimiano and Magnini (2005), this exploration is a relevant step in order to build ontologies in specialized domains. In line with this idea, Auger and Barrière (2010), Bodenreider (2001), as well as Ananiadou and McNaught (2006) explain the importance of terminological extraction as a previous phase for building ontologies, particularly in biomedicine given a certain level of complexity for distinguishing good candidates to terms.

Smith (2003) points out that there are two basic and universal lexical-semantic relations among concepts: Hyponymy-Hyperonymy and Part-Whole relations.

Olga Acosta, César Aguilar: Deparment of Language Sciences, Pontificia Universidad Católica de Chile, Av. Vicuña Mackenna 4860, Macul, Santiago, Chile, e-mail: {oacostal, caguilara}@uc.cl, http://cesaraguilar.weebly.com

In the first case, the works of Hearts (1992); Ryu and Choy (2005); Pantel and Pennacchiotti (2006); Snow, Jurafsky and Ng (2006); as well Ritter, Soderland and Etzioni (2009) among others, offer a relevant scope about the results obtained using hybrid methods. Such methods are based on the delimitation of specific lexical patterns derived from the operator *X* ***is_a*** *Y*, and the implementation of stochastic methods for evaluating the performance of the search systems designed for solving this task. In the second case, the experiments of Berland and Charniak (1999) and Girju, Badulescu and Moldovan (2006) have provided a set of patterns and methods for its automatic identification in large-text corpora in English. Additionally, Soler and Alcina (2010) have developed an experiment for extracting terms involved in this relation in a technical corpus in Spanish about ceramics.

Girju, Badulescu and Moldovan (2006) conceive the meronymy relations, also called *Part-Whole*, as links established between all units that integrate a set. In a similar way, Smith (2004) considers the part-whole relation is a relation of instances according to the following terms: the instances that belong to *A* are parts of instances that also belong to *B*, for example: *nucleus* ***part_of*** *cell*. If an instance of *nucleus* exists, then an instance of *cell* exists at the same time.

In line with these works focused to extraction of meronyms, we propose an experiment for recognizing concrete entities inserted in *spatial scenes*. Then, such concrete entities are linked with other concrete entities through the Spanish preposition *de* (Eng.: *of*). We assume that these physical entities related with *de* represent a part-whole relation. According to the taxonomy of Winston, Chaffin and Herrmann (1987), given that we focused on concrete entities the relations of our interest are mostly component-integral object and stuff-object. Additionally, we have in mind the experiments made for Acosta, Sierra and Aguilar (2011), as well Acosta, Aguilar and Sierra (2013), identifying relational adjectives for removing *noise* in NPs where concrete entities can be found.

The paper is organized as follows: in the section 2 we present a brief state of art about previous works oriented to the automatic detection of meronyms in general and specialized corpora. In section 3 we delineate our theoretical framework, particularly assuming a cognitive point of view respect to the configuration of spatial scenes derived from axial properties linguistically associated with *de*. In the section 4 we develop our method of extraction, based on the construction of a chunk grammar for detecting NPs with adverbs and adjectives linked to *de*, considering these phrases as candidates of terms that refer to concrete entities inserted in a Part-Whole relation. In section 5, we establish a second step in our methodology for removing non-relevant modifiers from candidates NPs. In the section 6 we show our results, and in section 7 we sketch our preliminary conclusions.

2 State of Art on Extraction of Part-Whole Relations

F Girju, Badulescu and Moldovan (2006), meronymy relations are an important phenomenon studied by philosophy, cognitive psychology and linguistics. In the case of computational linguistics, the works of Berland and Charniak (1999) as well as Girju, Badulescu and Moldovan (2006) have established two important methods for recognizing meronyms on corpora, combined with the use of lexical patterns and statistics. In the case of Berland and Charniak, they focused on genitive patterns identified through the employ of lexical seeds (*book, building, car, hospital, plant* and *school*) inserted in a part-whole relation with other words (e.g.: *the **basement** of a **building***). For this search, they use a news corpus of 100,000,000 words, and generate an ordered list of part-whole candidates inferred by a log-likelihood metric. Berland and Charniak obtained a level of accuracy around 55% respects to the words associated in a relation part-whole with such seeds. Once achieved these results, the authors compare them with the meronyms associated by WordNet (Fellbaum 1998) in order to determine precision.

In contrast, Girju, Badulescu and Moldovan developed another method that considers a large list (around 50) of possible patterns based on the relations considered by Winston, Chaffin and Herrmann (1987). Girju, Badulescu and Moldovan built a corpus conformed by texts obtained of LA Times and Wall Street Journal (WSJ). For detecting these patterns, they designed an algorithm named *Iterative Semantic Specialization Learning* (or ISSL). ISSL introduces a process of machine learning for recognizing new meronymy sequences in corpus, derived from 50 patterns mentioned. Such process needs a list of positive and negative examples for discerning good from bad candidates. In order to distinguish good candidates from bad candidates, the authors implemented a decision tree model which allows inferring rules of formation of meronymy sequences. In sum, ISSL achieves a level of precision almost 83% in the LA Times corpus, and 79% in the WSJ. In contrast, the level of recall is 79% for the first corpus, and 85% for the second corpus. Finally, they compare the meronymy relations identified with a set of meronyms from WordNet (Fellbaum 1998), in order to determine the degree of precision of their results.

In our case, our method explores a cognitive point of view in order to identify physical entities potentially involved in part-whole relations within a domain-specific corpus. According to Girju, Badulescu and Moldovan, in English 69% of the patterns of meronymy are expressed by the preposition *of*, genitives, the verb *have*, and noun compounds. In Spanish, genitives are only represented by preposition *de*, noun compounds have an explicit preposition, and the preposition *de* has an occurrence as prepositional modifier of a noun head over 85% in real corpus (data extracted from engineer corpus CLI and a domain-medical corpus). If we take into account the above mentioned data, our method focus on extracting part-whole instances of the most common cluster. To achieve our goal, we only consider specific elements as place adverbs, as well as the most frequent patterns of building of terms in order to identify physical entities. Furthermore, linguistic heuristics are used for recognizing and removing descriptive adjectives from NPs with the goal of extracting the best candidates to concrete entities.

3 Theoretical Framework

We propose a method for extracting part-whole relations where concrete entities are involved. We assumed that physical entities participate in *spatial scenes* as *reference objects*. *Spatial scenes* are configured around *axial properties* mapped by adverbs linked to the preposition *de* (Eng.: *of*). In the following sub-sections, we establish specific distinctions between *axial properties* and *spatial scenes*.

3.1 Axial Properties

According to Evans (2007), a *spatial scene* is a linguistic unit that contains information based on our spatial experience. Such space is structured according to four parameters: a *figure* (or *trajector*), a *referent object* (that is, a *landmark*), a *region* and – in certain cases – a *secondary reference object*. These two reference objects configure a *reference frame*. We can understand this configuration analyzing the following example: *A car is parked behind the school*. In this sentence, *a car* is the *figure* and *the school* is the *reference object*. Respect to the *region*, this is established by the combination of the preposition which sketches a spatial relation with the referent object. Finally, such relation encodes the location of the *figure*.

On the other hand, Evans (2007) points out the existence of *axial properties*, that is, a set of spatial features associated to a specific *reference object*. Considering again the sentence *a car is parked near to the school*, we can identify the location of the car searching for it in the region near to the *school*. Therefore, this search can be performed because the referent object (*the school*) has a set of *axial divisions*: front, back and side areas. These *axial properties* configure all spatial relations.

3.2 Axial Properties and Place Adverbs

Axial properties are linguistically represented by place adverbs. In this experiment we only consider adverbs functioning in Spanish with *de*, e.g.: *enfrente de* (*in front to/of*), *detrás de* (*behind*), and so on. *De* is one of the most used prepositions for representing many relations, particularly, part-whole relations (Berland and Charniak, 1999; Girju, Badulescu and Moldovan, 2006). Table 1 shows place adverbs considered in our experiment. Additionally, we use some nouns such as exterior and interior that they are related with adverb *in*.

Tab. 1: Place adverbs representing Spatial Relations.

Adverb in Spanish
Enfrente, delante (Engl. *in front to/of*)
Detrás (Engl. *behind*)
Sobre, encima (Engl. *on*)
Abajo, debajo (Engl. *under*)
Dentro (Engl. *in*)
Arriba (Engl. *above, over*)
Lado (Engl. *beside, near*)

3.3 Adjectives

If we take into account the internal structure of adjectives, two kinds of adjectives can be identified: permanent and episodic adjectives (Demonte, 1999). The first type of adjective represents stable situations, permanent properties characterizing individuals. These adjectives are located outside of any spatial or temporal restriction. On the other hand, episodic adjectives refer to transient situations or properties implying change and with a space-temporal limitation. Almost all descriptive adjectives derived of participles belong to this latter class as well as all adjectival participles.

The Spanish language is one of the few languages that represent in the syntax this difference in the meaning of the adjectives. Such difference in many languages is only recognizable by means of the interpretation. In Spanish, individual properties can be predicated with the verb *ser*, and episodic properties with the verb *estar*, which is an essential test to recognize what class an adjective belongs to.

4 Methodology

In this work we propose a methodology for recognizing concrete entities from text sources. Then, such entities are related with other concrete entities by means of the preposition *de*. In the following subsections we develop a description of each phase of our methodology.

4.1 Part-Of-Speech Tagging

Part-of-speech tagging (POS) is the process of assigning a grammatical category or part of speech to each word in a corpus. For instance, one POS output can provide the word form, the part-of-speech tag and the lemma with the next structure:

Word form	**Tag**	**Lemma**
Defined	VM	Define

The next example shows a sentence in Spanish tagged with the FreeLing tagger (Carreras et al. 2004):

> el/DA tipo/NC más/RG común/AQ de/SP lesión/NC ocurrir/VM cuando/CS algo/PI irritar/VM el/DA superficie/NC externo/AQ del/PDEL ojo/NC

Tags used by FreeLing are based on the tags proposed by the EAGLES group for the morphosyntactic tagging of lexicons and corpus for all European languages. Methodology proposed in this work requires POS because we focused on regular patterns of terms and place adverbs representing *axial properties* linked with a *reference object* by means of preposition *de*.

4.2 Standardizing POS

In this work we used FreeLing tool to tag text with POS. With the goal of standardizing the use of preposition *of* in both cases contraction (del/SPC) and preposition plus article (de/SP el/DA), we replace both cases with del/PDEL tag. Similarly, in most cases tags were reduced to two characters except in the case of *to be* verb (i.e., VAE).

4.3 Chunking

Chunking is the process of identifying and classifying segments of a sentence by means of the grouping of major parts-of-speech forming basic non-recursive phrases. A set of rules indicating how sentences should be grouped makes up a grammar called *chunk grammar*. The rules of a chunk grammar use tag patterns to describe sequences of tagged words, e.g. <da>?<qa>*<nc>+. Tag patterns are similar to regular expression patterns, where symbols such as "*" means zero or more occurrences, "+" means one or more occurrence and "?" represents an optional element.

In this work, we are interested in extracting concrete entities from text information. A concrete entity can be a primary *reference object* implicit in a *spatial scene*. Linguistically, a *spatial scene* can be a sentence as: *the bike is in front of the house*. In the above example, we can locate a *bike* (e.g., the figure) by looking for *axial properties* of the *house*. As mentioned above, linguistically, *axial properties* can be represented by place adverbs. Therefore, in this first experiment we propose the following regular pattern for extracting spatial scenes where a concrete entity (reference object) can be implicit:

(<NC>+<AQ>*<VAE>)?<RG><PDEL><DA>?<NC><AQ>* (1)

From above pattern, we have an optional structure that can be the *figure*, with a specific syntactic structure (at least one common noun and optional adjectives), and *to be* verb. Then, there is an obligatory adverb indicating *axial properties* of a *reference object*. Finally, the pattern considers a *reference object* with a similar structure to *figure*. *Figure* structure and *to be* verb are considered optional because this allow us to increase recall.

On the other hand, structure of NPs for bootstrapping phases in order to extract other entities related with the reference object is:

<NC >+<AQ>*<PDEL><NC><AQ>* (2)

5 Removing Noise

5.1 Obtaining Non-Relevant Adjectives

In the framework of the extraction of definitional contexts, Acosta *et al.*, (2011) based on Demonte (1999) implemented a series of linguistic heuristics for the extraction of descriptive adjectives in order to remove *noise* in results. These heuristics outperformed information theory measures such as pointwise mutual information (PMI) for filtering relevant adjectives co-occurring with a potential *Genus Term*. In this experiment, we consider one of the heuristics implemented by Acosta et al., (2011): an adverb preceding an adjective, and another new heuristic where *estar* verb precedes an adjective. This latter as mentioned in previous sections on the distinction between individual and episodic adjective:

<RG><AQ> (3)

<VAE><AQ> (4)

5.2 Obtaining Non-Relevant Nouns

As a reference corpus in order to remove non-relevant nouns, we used a collection of Spanish press articles with an extension of two million words downloaded from the Leipzig Corpora Collection (Goldhahn, Eckart and Quasthoff, 2012). We filtered nouns by means of setting a threshold of occurrence of nouns in general corpus.

5.3 Removing Non-Relevant Nouns and Adjectives from Phrases

Before extracting concrete entities, the non-relevant set of automatically obtained nouns and adjectives are removed from the set of the phrases:

<NC>+<AQ>*<PDEL><NC><AQ>* (5)

(<NC>+<AQ>*<VAE>)?<RG><PDEL><DA>?<NC><AQ>* (6)

5.4 Extracting Part-Whole Relations

Winston, Chaffin and Herrmann (1987) proposed a taxonomy of part-whole relations. According to this taxonomy, our methodology focuses on the extraction of relations component-integral object and stuff-object. These relations can be identified by linking potential concrete entities with other elements from the same specialized corpus. This link is done by means of preposition *of*. For instance, if *hand* is extracted as a concrete entity, then, we can identify all of elements linked with *hand* by means of the preposition *of* (*finger of the hand*, *skin of the hand*, and so on).

6 Results

6.1 Sources of Textual Information

The source of textual information is constituted by a set of documents of the medical domain, basically human body diseases and related topics (surgeries, treatments, and so on). These documents were collected from MedLinePlus in Spanish. MedlinePlus is a site with a goal to provide information about diseases, treatments, and conditions that are easy to understand. The kind of communication used in this textual source can be considered as expert-beginner because the information is intended for patients, families, and it is created by various health institutes. Taking into account that each knowledge area has a different lexical set, this kind of communicative situation will be most explanatory using definitions where the meaning of the lexical set must be clarified.

The size of the corpus is 1.2 million of words. We chose a medical domain for reasons of availability of textual resources in digital format. Furthermore, we assume that the choice of this domain does not suppose a very strong constraint for generalization of results to other domains.

6.2 Other Resources

The programming language used in order to automate all tasks required was Python as made available in the NLTK module. NLTK module constitutes a very valuable resource for research and development in natural language processing. On the other hand, POS tagger used in this experiment was FreeLing.

6.3 Analysis of Results

The first phase of extraction of concrete entities had a high precision of 73%. In a next *bootstrapping* step precision decreased to 51%. Next phases (3 to 5) hold measures without major changes in precision. In relation to the candidate Part-Whole relations were only extracted with the set obtained of the phase 1. Given the low precision obtained in phase 1 (24%), candidates of remaining phases were not analyzed. In general terms, recall achieved in the extraction of concrete entities was 65%.

We considered that one of the relevant results of our experiment, which was not addressed in the initial design, is the precision in recognition of terms of the domain. As it can be observed in the table 2, precision is over a 70%, which is a relevant result. On the other hand, we noted extraction of concrete entities tend to converge in a phase 5.

Tab. 2: Results obtained.

Phase	Precision	Concrete entities	Candidates	Terms
1	73%	308	423	74%
2	51%	566	1109	74%
3	46%	744	1632	72%
4	44%	806	1849	71%
5	43%	821	1891	71%

7 Conclusions

Extraction of lexical relations is a very relevant task in Natural Language Processing. Given the huge amounts of text information available, the possibility of organizing a conceptual space in an automatic way justifies any effort aimed at this purpose. In this work we take into account research conducted in cognitive linguistics about spatial scenes in order to extract relevant information related to concrete entities. These concrete entities were used for identifying and extracting part-whole relations. Although precision in the extraction of part-whole relations was low, we consider that it can be improved by other filtering methods that we will carry out in the future.

Finally, one of the relevant results of our experiment was the precision in the recognition of terms. As data show, over 70% of the elements obtained were terms, which can be considered a relevant result. These results were not addressed in the original design of the experiment, but we now consider that the extraction of physical entities could be an important route for terminology extraction because this set could be considered as a good starting point for building new terms. For example, single nouns could be linked to relational adjectives or prepositional phrases mostly preposition *de* for finding new terms.

Acknowledgments

This paper has been supported by the National Commission for Scientific and Technological Research (CONICYT) of Chile, Project Numbers: 3140332 and 11130565.

Bibliography

Acosta, O., Sierra, G. And Aguilar, C. (2011) Extraction of Definitional Contexts using Lexical Relations. *International Journal of Computer Applications*, 34(6): 46–53.

Acosta, O., Aguilar, C. and Sierra, G. (2013) Using Relational Adjectives for Extracting Hyponyms from Medical Texts. In: Lieto, A. and Cruciani, M. (eds.), *Proceedings of the First International Workshop on Artificial Intelligence and Cognition (AIC 2013)*, CEUR Workshop Proceedings, Torino, Italy, pp. 33–44.

Ananiadou, S. and McNaught, J. (2006) *Text Mining for Biology and Biomedicine*. Artech House, London.

Auger, A. and Barrière, C. (2010) Probing *Semantic Relations: Exploration and Identification in Specialized Texts*. John Benjamins Publishing, Amsterdam/Philadelphia.

Berland, M. and Charniak, E. (1999) Finding parts in very large corpora. In: *Proceedings of the 37th Annual Meeting of the Association for Computational Linguistics on Computational Linguistics*. Association for Computational Linguistics, pp. 57–64.

Bodenreider, O. (2001) *Medical Ontology Research: A Report to the Board of Scientific Counselors of the Lister Hill National Center for Biomedical Communications*. National Institutes of Health, Bethesda, Maryland, USA.

Buitelaar, P., Cimiano, P. and Magnini, B. (2005) *Ontology learning from text.* IOS Press, Amsterdam.

Carreras, X., Chao, I., Padró, L. and Padró, M. (2004) FreeLing: An Open-Source Suite of Language Analyzers. In: *Proceedings of the 4th International Conference on Language Resources and Evaluation LREC 2004*. ELRA Publications, Lisbon, Portugal, pp. 239–242.

Demonte, V. (1999). El adjetivo. Clases y usos. La posición del adjetivo en el sintagma nominal. In: *Gramática descriptiva de la lengua española*, Vol. 1, Cap. 3, pp. 129–215.

Evans, V. (2007) *A Glossary of Cognitive Linguistics*. Edinburgh University Press, Edinburgh, UK.

Girju, R., Badulescu, A. and Moldovan, D. (2006) Automatic discovery of part–whole relations. *Computational Linguistics*, 32(1), 83–135.

Goldhahn, D. Eckart, Th. and Quasthoff, U. (2012) Building Large Monolingual Dictionaries at the Leipzig Corpora Collection: From 100 to 200 Languages. In: *Proceedings of the Eight International Conference on Language Resources and Evaluation LREC'12*, ELRA Publications, Istanbul, Turkey, pp. 759–765.

Hearts, M. (1992) Automatic acquisition of hyponyms from large text corpora. In: *Proceedings of Conference COLING*. Nantes: Association for Computational Linguistics.

Pantel, P. and Pennacchiotti, M. (2006) Espresso: Leveraging Generic Patterns for Automatically Harvesting Semantic Relations. In: *21st International Conference on Computational Linguistics & 44th annual meeting of the Association for Computational Linguistics*, Sydney, Australia, pp. 113–120.

Ritter, A., Soderland, S., and Etzioni, O. (2009) What is This, Anyway: Automatic Hypernym Discovery. In: *Papers from the AAAI Spring Symposium*, Menlo Park, Cal., AAAI Press, pp. 88–93.

Ryu, K., and Choy, P. (2005) An Information-Theoretic Approach to Taxonomy Extraction for Ontology Learning. In: Buitelaar, P., Cimiano, P., and Magnini, B. (eds.), *Ontology Learning from Text: Methods, Evaluation and Applications*, IOS Press, Amsterdam, pp. 15–28.

Smith, B. 2003. Ontology. In: Floridi, L. (ed.), *Blackwell Guide to the Philosophy of Computing and Information*. Blackwell. Oxford, pp. 155–166.

Smith, B. (2004) Beyond concepts: ontology as reality representation. In: Varzi, A. and Vieu, L. (eds.), *Formal Ontology and Information Systems*, IOS Press, Amsterdam, pp. 73–84.

Snow, R., Jurafsky, D. and Ng, A. (2006) Semantic taxonomy induction from heterogeneous evidence. In: *Proceedings of the 21st International Conference on Computational Linguistics*. Sydney, Australia, Association for Computational Linguistics.

Soler, V. and Alcina, A. (2010) Patrones léxicos para la extracción de conceptos vinculados por la relación parte-todo en español. In: Auger A. and Barrière, C. (eds.), *Probing Semantic Relations: Exploration and Identification in Specialized Texts*, John Benjamins Publishing, Amsterdam/Philadelphia, pp. 97–120.

Winston, M., Chaffin, R. and Herrmann, D. (1987) A taxonomy of part-whole relations. *Cognitive science* 11(4): 417–444.

Wiesław Lubaszewski, Izabela Gatkowska, and Marcin Haręza

Human Association Network and Text Collection

A Network Extraction Driven by a Text-based Stimulus Word

Abstract: According to tradition, experimentally obtained human associations are analyzed in themselves, without relation to other linguistic data. In rare cases, human associations are used as the norm to evaluate the performance of algorithms, which generate associations on the basis of text corpora. This paper will describe a mechanical procedure to investigate how a word embedded in a text context may select associations in an experimentally built human association network. The results are rather encouraging: this procedure is able to distinguish those words in a text which enter into a direct semantic relationship with the stimulus word used in the experiment to create a network, and is able to separate those words of the text which enter into an indirect semantic relationship with the stimulus word.

1 Introduction

It is easy to observe that semantic information may occur in human communication, which is not lexically present in a sentence. Consider, for example, this exchange: *Auntie, I've got a terrier! – That's really nice, but you'll have to take care of the animal.* The connection between the two sentences in this exchange suggests that there is a link between *terrier* and *animal* in human memory. It is well known that it is possible to investigate such connections experimentally (Kent, Rosanoff 1910). The free word association test, in which the tested person responds with a word associated to a stimulus word provided by an investigator, would create a network in which words are linked with multiple links (Kiss et al. 1973). The links which associate a specific word to other words are supposed to define the meaning of this specific word.

Wiesław Lubaszewski, Izabela Gatkowska: Jagiellonian University, Gołębia 24, 31-007 Kraków, Poland, e-mail: lubaszew@agh.edu.pl, izabela.gatkowska@uj.edu.pl, www.klk.uj.edu.pl
Marcin Haręza: Computer Science Department, AGH University of Science and Technology, Al. Mickiewicza 30, 30-059 Kraków, Poland, e-mail: mhareza@student.agh.edu.pl

The experimentally built network gives an opportunity to investigate how a word embedded in a text context refers to a network. This investigation may be based on experiments involving human participants. But the experimentally built network is a really large structure. Then it seems to be reasonable to look for mechanical procedures which would help the investigation.

This paper describes a mechanic procedure which investigates how a word of a text may use context words to select associations in human association network. Technically speaking, the procedure is an algorithm, which would identify in a text the stimulus word and the association words directly related to it, then, using those words, would find in the net, the sub-net (sub-graph) which is optimally related to the stimulus word – where optimal means a sub-net in which each node (word) is semantically related to a stimulus word. The paper will also describe the tests of the procedure performed on a three different text collections and will provide a detailed evaluation of the results.

2 The Network

The original word association experiment was designed to produce an association norm, and therefore it was a single phase, i.e. a test in which associations were not deliberately used as stimuli. To built a rich association network which would explain indirect associations, e.g. *owca* 'sheep' – *wełna* 'wool', where there is no explicit and direct semantic relation between the stimulus and the response, we have to redesign the original experiment into a multiphase test, in which associations from phase one would serve as stimuli in phase two, and so on (Kiss et al. 1973). The resulting network is a structure built from lexical nodes and relations.

The network described in this paper was built via a free word association experiment (Gatkowska 2014) in which two sets of stimuli were employed, each in a different phase of the experiment. In the first phase, 62 words taken from the Kent-Rosanoff list were tested as the primary stimuli. In the second phase, the 5 most frequent responses to each primary stimulus obtained in phase one were used as stimuli. To reduce the amount of manual labor required to evaluate the algorithm output, we used a reduced network, which is based on:
- 43 primary stimuli taken from the Polish version of the Kent-Rosanoff list.
- 126 secondary stimuli which are the 3 most frequent associations to each primary stimulus.

The average number of associations to a particular stimulus produced by 540 subjects is about 150. Then the total number of stimulus–response pairs obtained for the 168 stimuli as a result of the experiment is equal to 25,200. Due to the fact that the analysis of the results produced by an algorithm would require manual work, we reduced the association set through the exclusion of each stimulus–response pair, where the response frequency was equal to 1. As a result, we obtained 6,342 stimulus–response pairs, where 2,169 pairs contain responses to the primary stimuli, i.e. primary associations, and 4,173 pairs which contain responses to the secondary stimuli, i.e. secondary associations. The resulting network consists of 3.185 nodes (words) and 6155 connections between nodes.

The experimentally built association network can be described on a graph, where the *graph* is defined as tuple (V, E), where V is the set of nodes (vertices) and E is the set of connections between the two nodes (vertices) from V. As each stimulus–association (response) pair has a direction, an association network is a *directed* graph (Kiss et al. 1973), which means that each connection between the two nodes $(v1, v2)$ has a direction, i.e. it starts in $v1$ and ends in $v2$. If one recognizes that the connection $(v1, v2)$ is a semantic relation between the meanings of the two words, then one has to distinguish the two types of connections. First are connections such as *chair – leg* or *leg – chair*, where the stimulus–response direction and the direction of semantic relation between the two meanings is the same. Second are connections such as *chair – table* or *table – chair*, where the direction of the semantic relation and the stimulus–response direction may differ. Therefore one can treat the association network as an *undirected* graph, which means that the connection between the two nodes $(v1, v2)$ has no direction, i.e. $(v1, v2) = (v2, v1)$: this kind of connection is called an *edge*. The connection between the two nodes may have a *weight*. The experiment result is a list of triples: (S, A, C), where S is the stimulus, A is the association and C is the number of participants, who are associated A to S. The C represents the association strength, which can be converted into a connection weight of C_w, counted as follows: $C_w = S_c/C$, where S_c is the total of all responses given to the stimulus S. Then we may treat the association network as an *undirected weighted* graph, which is a tuple (V, E, w), where w is the function that assigns every edge a weight. Then the *path* in the graph is a sequence of nodes that are connected by edges. The path *length* is the number of nodes along the path. Path *weight* is the sum of the weights of the edges in the path. The *shortest* path between two nodes $(v1, v2)$ is the path with the smallest path weight.

3 The Network Extraction Driven by a Text-Based Stimulus

If both the network and the text are structures built of words, then we may look for an efficient algorithm that can identify in the text the primary stimulus-word and a reasonable number of primary associations to this stimulus. Words identified in the text may serve as the starting point to extract from the network a sub-graph, which will contain as many primary and secondary associations as possible. The semantic relationship between the nodes of a returned sub-graph will be the subject of evaluation.

In more technical language, the algorithm should take a graph (association network) and the subset of its nodes identified in a text (*extracting nodes*) as an input. Then the algorithm creates a sub-graph with all extracting nodes as an initial node set. After that, all the edges between extracting nodes which exist in the network are added to the resulting sub-graph – these edges are called *direct* ones. Finally, every direct edge is checked in the network, to find whether it can be replaced with a shorter path, i.e. path which has a path weight lower than the weight of the direct edge and has a node number smaller than or equal to the predefined length. If such a path is found, it is added to the sub-graph – where add means adding all the path's nodes and edges.

3.1 Extraction Procedure

The source graph *G*, extracting nodes *EN* and maximum number of intermediate nodes in path *l* are given. First, an empty sub-graph, *SG*, is created, and all extracting nodes, *EN*, are added to a set of vertices V_{sg}. In next step set, the *ENP* of all pairs between the vertices in the *EN* is created. For every pair in the *ENP* algorithm checks, if the edge between the paired vertices $v1, v2$ exists in *G*. If it does, this edge is added to the sub-graph *SG* set of edges E_{sg}. Then the shortest path sp between $v1$ and $v2$ is checked in *G*. If the shortest path *sp* is found, i.e. *sp* weight is lower than the direct edge $(v1, v2)$ and the number of path intermediate nodes is less than or equal to *l*(*length*(*sp*) – 2, (-2 because of the start and end nodes are not intermediate), then the *sp* path is added to the sub-graph *SG* by adding its vertices and edges to the appropriate sets V_{sg} and E_{sg}. Finally, the sub-graph *SG* is returned.

It seems to be clear that the size of the sub-graph created by the algorithm depends on the number of extracting nodes given on the input. As the texts may differ in the number of primary associations to a particular stimulus which would

Input: $G - (V, E, w)$ – graph (set of vertices, set of edges, weighting function), EN – extraction nodes, l – maximal number of intermediate vertices in the path
Output: $SG - (V_{sg}, E_{sg}, w_{sg})$ – extracted subgraph

```
V_sg ⟵ V_g;
E_sg ⟵ ∅;
w_sg ⟵ w_g;
ENP ⟵ pairs(EN);
foreach v1, v2 ∈ ENP do
    if edge(v1, v2) ∈ G then
        E_sg ⟵ E_sg + edge(v1,v2);
        sp ⟵ shortest_path(v1,v2);
        if weight(sp) < w_g(edge(v1,v2)) and length(sp) −2 ≤ l then
            foreach v ∈ vertices(sp) do
                if v ∉ V_sg then
                    V_sg ⟵ V_sg + v;
                end
            end
            foreach e ∈ edges(sp) do
                if e ∉ E_sg then
                    E_sg ⟵ E_sg + e;
                end
            end
        end
    end
end
return SG
```

Algorithm 1: *EA(G, EN, l).*

serve as extracting nodes, there is a need for a procedure that would control the number of extracting nodes used by the network-extracting algorithm.

3.2 The Control Procedure

To be used for building a sub-graph for a given stimulus *S*, the text must contain at least *RPn* primary associations to an *S* stimulus. Choosing the proper *RPn* could be done manually, but an algorithm which adapts the *RPn* to a particular text and particular stimulus is introduced. The $RPn = 2$ is chosen as starting value for this algorithm. If the text has $RPn < 2$, the text is omitted. If the text has $RPn \geq 2$, the text is used for sub-graph extraction. First, the stimulus and $RPn = 2$ primary associations are passed as extracting nodes to the network extracting algorithm *NEA*. Then the number of nodes in the returned sub-graph is counted. Next, the *RPn* is incremented by 1 and the new sub-graph is created. Finally, the number of nodes of the sub-graph based on $RPn + 1$ is multiplied by the P parameter, which says which fraction of the base sub-graph (created for $RPn = 2$) must exist in the final sub-graph – for example $P = 0.5$ means that at least half of the nodes from the base sub-graph must remain in the sub-graph created after the *RPn* adaptation. If the newly created sub-graph does not match the condition set by *P*, the *RPn* value is decremented by 1 and becomes the final *RPn* value. If the newly created sub-graph matches the condition set by *P*, the *RPn* is incremented by 1 and the new sub-graph is created.

4 Tests of the Network Extracting Procedure

4.1 The Corpora to Perform Tests

In order to compare the association network with the texts, we have used the three stylistically and thematically different corpora previously described (Gatkowska et al. 2013). All three corpora were lemmatized using a dictionary based approach (Korzycki 2012).

The first corpus consists of 51,574 press communiques of the Polish Press Agency and contains over 2,900,000 words. This naturally randomized corpus represents a very broad variety of real life events described in a simple, informative style. This corpus will be referred to as PAP.

The second corpus is a fragment of the National Corpus of Polish with a size of 3,363 separate documents (Przepiórkowski et al. 2011) spanning over 860,000 words. This corpus was intended to be the representative one, i.e. one which covers all Polish language phenomena. This corpus will be referred to as the NCP.

The last corpus is composed of 10 short stories and the novel *Lalka* (The Doll) by Bolesław Prus – a late XIX century novelist using a version of modern Polish similar to present-day varieties. The texts are split into 10,346 paragraphs of over 300,000 words. The rationale behind this corpus was to match the association network to the texts, which were intentionally written to interact with the reader's imagination. This corpus will be referred to as Prus.

4.2 Corpus-Based Primary Stimulus Sub-Graph Extraction

First, a separate sub-graph for each primary stimulus was created for each text in the corpus. All sub-graphs were obtained with empirically adjusted parameters (Haręza 2014), such as: intermediate nodes in the path of $l = 3$ for the extracting algorithm, and primary associations to stimulus minimum $RPn = 2$, with a sub-graph adjusting parameter of $P = 0.5$ for the control procedure. Then the text based sub-graphs obtained for a specific primary stimulus were merged into the corpus based primary stimulus sub-graph, i.e. all sets of nodes and all sets of edges were merged, forming a multiple set union. Finally, the corpus-based primary stimulus sub-graph was trimmed, which means that each node which does not match the T parameter was removed from the sub-graph. The T parameter sets the maximum number of edges between primary stimulus and the node in the sub-graph. To match the network-forming principle that a primary stimulus produces a primary association, which then serves as a secondary stimulus to produce a secondary association, the $T = 2$. The corpus based primary stimulus sub-graph was extracted separately for each corpus described above.

4.3 Evaluation of Extracted Sub-Graph Quality

Preliminary Evaluation. To perform the evaluation, each of the 6.342 stimulus–response pairs used to build the network was manually evaluated. If the stimulus is semantically related to the response as in *dom – ściana* (house – wall), the pair is marked as positive, otherwise the pair is considered to be a negative one, e.g. *góra – Tatry* (mountain – the mountains) or *dom – własny* (house – own).

Then the sub-graph nodes were evaluated consecutively along a path in the following way. If the two connected nodes matched a positive stimulus–response pair, then the right node was marked as positive – *pNode*. If the two connected nodes matched a negative stimulus–response pair, then the right node was marked as negative one – *nNode*. If the two connected nodes did not matched any stimulus–response pair, then both nodes were marked as negative. The primary stimulus node is on principle a positive one.

Sub-graph quality was evaluated for each of the 43 primary stimuli separately for each corpus. Then the corpus-based average score for the 43 sub-graphs was calculated by summing up the positive and negative nodes in the sub-graph. The result is in Table 1.

The results show that the best proportion between negative and positive nodes is observed in the sub-graph based on the Prus corpus (0.0098 compared to NCP 0.027 and PAP 0.030), but we shall not discuss the differences between

Tab. 1: Sub-graph Nodes Evaluated for 3 Corpora.

NCP		PAP		Prus	
pNod	nNod	pNod	nNod	pNod	nNod
783	27	898	27	406	4

the corpora here. We decided that our source for detailed corpus based sub-graph evaluation will be the PAP corpus, which was built from complete texts, and which was the basis for the creation of the sub-graphs, which contain the highest number of positive nodes.

Rigorous Evaluation. Now we can present the results for each primary stimulus, where the sub-graph for each primary stimulus word is evaluated. To perform this evaluation, we shall compare a sub-graph nodes with the texts which were the base for sub-graph extracting. After then we shall compare the network, the sub-graph and the texts, to find if there are nodes which are present both in the network and in the texts but not in the sub-graph. Finally, we shall perform a rigorous evaluation of the semantic consistency of the sub-graph. The results of the comparison are presented as:

SnT – number of primary and secondary association nodes which were recognized both in the texts and in the sub-graph

TNn – number of primary and secondary semantic associations nodes which are present in the texts and in the network but were rejected by the algorithm and therefore they are not present in the sub-graph.

Sn – number of nodes in the sub-graph created by the algorithm,

NnS – number of negative nodes in the sub-graph recognized by a manual evaluation. In this case the evaluation was more rigorous than the preliminary one, which means that each node in the sub-graph was tested against the primary stimulus node. A negative node is considered to be any node (association) which does not enter into a semantic relation to the primary stimulus, even if it enters into a semantic relation with a secondary stimulus node, e.g. the path: *krzesło* 'chair' – *stół* 'table' – *szwedzki* 'Swedish', where pairs *krzesło* – *stół* and *stół* – *szwedzki* enter into a semantic relation, but the primary stimulus *krzesło* 'chair' does not enter into a semantic relationship with secondary association *szwedzki* 'Swedish'.

Before we start the evaluation per stimulus, we have to show the joint results of the evaluation of 43 stimuli. To perform this analysis we must know the number of nodes in the network – *Nn*. Table 2 shows the joint result for all sub-graphs based on the PAP corpus.

Tab. 2: Joint Evaluation of 43 Stimuli.

Prim. Stimulus	*Nn*	*SnT*	*Sn*	*NnS*	*TNn*
43	3185	710	898	65	38

If one looks at Table 2 and compares the numbers of network nodes *Nn* and sub-graph nodes retrieved in text *SnT*, one can see that only a fraction (0.22) of the words (nodes) which are present in the network appear in the large text collection. The score is substantially lower then the sub-graph node *Sn* to network node ratio, which is 0.28. It can be said that those figures show the relation between the language system (the network) and the use of language (the texts).

The *NsS* value (negative nodes in the sub-graph) shows that the negative nodes in sub-graph are only 0.72 of the total sub-graph nodes. If one compares this score with the negative nodes ratio presented in table 1 (0.30), it is possible to say that a restrictive analysis which is based on the semantic relatedness of the primary stimuli and association to the secondary stimuli more than doubles the negative nodes in the sub-graph, but the negative nodes are still below the 1 percent of the sub-graph nodes. This result indicates that the cautious method for building a sub-graph provides a good output, and that the method described in this paper can be treated as reliable.

Finally, the *TNn* compared to the *ANn* show that the associations semantically related to the primary stimulus, which are present both in the network and the texts, but were excluded from the sub-graph, are 0.053 of network nodes, but we shall provide a detailed analysis later on.

Results per Stimulus. A more detailed evaluation of the results will be possible if one takes a look at the results obtained for each particular primary stimulus. This results are shown in Table 3. Table includes also *PnN* (the number of primary association nodes in the network) to illustrate the stimulus status in the network.

At first glance, one may say that the stimuli status in the network seems to be similar, but the values of other parameters differ, and we can not relate those differences to the semantic features of the stimuli words. The differences in *SnT* (primary and secondary associations both in the sub-graph and the texts) seems to be random and corpus dependent – for example stimulus word *dywan* 'carpet' occurred only in 7 texts and only 2 of them were rich enough to provide the extracting nodes (at least 2 primary associations). The *SnT* is smaller than or equal to the *Sn* (sub-graph nodes) for most words – the only exceptions are: *żołnierz* 'soldier', *głowa* 'head', *ręka* 'hand', *radość* 'joy', and *pamięć* 'memory', but those stimuli words do not share a distinctive semantic feature.

Finally, we have to analyze the *TNn*, i.e. primary and secondary associations which are present both in the network and texts but are not present in the sub-graph. As this seems to be a real finding – all those associations are semantically related to a primary stimulus. But for most of them one can not explain their relation to a primary stimulus by a single semantic relation, e.g. *roof* `part_of` *house*. In the network almost all words qualified as TNn are related to a primary stimulus by a relation chain, e.g. relation *owca* 'sheep' – *rogi* 'horns' can be explained by the consecutive relations: *owca* 'sheep' `complementary_to` *baran* 'ram', followed by *baran* 'ram' `consits_of` *rogi* 'horns'. That is to say that all words qualified as *TNn*, relate to a primary stimulus in the same way as indirect associations (Gatkowska 2013). Therefore, the method described in this paper may automatically identify indirect associations, which are present in the network. We have to present all 12 *TNn* as found in tests:

żołnierz 'soldier'	*militarny* 'military'
owca 'sheep'	*rogi* 'horns', +*mięso* 'meat', **owieczka* 'baby sheep'
mięso 'meat'	*śniadanie* 'breakfast'
praca 'work'	*maszyna* 'machine', *decyzja* 'decision', *ręka* 'hand'
chłopiec 'boy'	*palec* 'finger'
woda 'water'	+*głębokość* 'depth' , *sól* 'salt', *piasek* 'sand', *fala* 'wave'
miasto 'water'	*rzeka* 'river', *przyroda* 'nature'
król 'king'	*prawo* 'law'
radość 'joy'	*serdeczny,* 'heartfelt' *strach* 'fear', *cieszyć się* 'enjoy', *rozczarowanie* 'disappointment', *rozpacz* 'despair', +*duży* 'big', *nieszczęście* 'disaster', *troska* 'worry', **radosny* 'joyfull'
pamięć 'memory'	*praca* 'work', *ocena* 'mark', *wola* 'will'
światło 'light'	**światłość* 'overwhelming light', *czytanie* 'reading', *pokój* 'room'
rzeka 'river'	*powietrze* 'air', *ziemia* 'earth'

The asterisk '*' marks all associations, which are not indirect semantic associations – in our results these associations are based on morphological relations between the stimulus and the association. The plus symbol '+' marks rare words, which enters into direct semantic relation to the primary stimulus.

Having compared the sub-graph, network and texts we can analyze the *NnS* – negative nodes in sub-graph. It can be seen that only 15 out of 43 sub-graphs does not contain a negative node. The number of negative nodes per primary stimulus differs –14 primary stimuli have a negative node of 1 in the sub-graph. For the last 14 primary stimuli negative nodes, the number differs from 2 to 8. This result was the reason for a manual check of each negative node. The analysis shows that each negative node is a word, whose meaning enters into a pragmatic relation to a primary stimulus word, where pragmatic mean association, which refers to a real world knowledge or to a particular real world situation. And in this moment, it is hard to explain, why *woda* 'water', *mężczyzna* 'male', *dziecko* 'child' and *góra* 'mountain/up' have the highest number of pragmatic associations in a sub-graph related to a primary stimulus. To summarize the *NnS* analysis, one can state that 29 out of 43 sub-graphs have negative node ratios much below an average of 0.72, but 14 out of 43 have a negative nodes ratio much higher than the average.

5 Brief Discussion of Results and Related Work

The proposed method for text-driven extraction of an association network is simple and cautious on graph operations. But the automatic comparison of extracted sub-graph with network and texts is able to distinguish in the texts those words which would enter into a indirect semantic relationship with the stimulus word used in the experiment. This means that the text-driven network extracting procedure may serve as a tool to provide the data which would locate the indirect association in a large network. This is a task which would be extremely hard to do manually.

On the other hand, the quality of sub-graphs extracted for words such as *pająk* 'spider', *lampa* 'lamp', *dywan* 'carpet', which occurred in a really small number of texts, seems to prove that the extracting algorithm does not depend on the number of texts used for network extracting. If it is true, the algorithm may serve as reliable tool for extracting an association network on the basis of a single text, which may provide data for the study of human comprehension – a human reader comprehends just the text, not a text collection. But this should be a subject of further investigation.

The problem of negative nodes in the sub-graph should be the subject of further investigation with the use of a richer network, i.e. a network that includes single responses.

Our premise that semantic information may occur, which is not lexically present in a text differs from the assumption "that semantic association is related to the textual co-occurrence of the stimulus – response pairs" (Schulte im Walde, Melinger 2008), then it is hard to compare our results. But there are possible some detailed observations.

Tab. 3: Rigorous Evaluation for each Primary Stimulus Word.

Prim. Stimulus	*PnN*	*SnT*	*Sn*	*NnS*	*TNn*
dywan 'carpet'	41	5	7	1	0
żołnierz 'soldier'	49	34	34	0	1
owca 'sheep'	54	19	19	2	3
ser 'cheese'	35	8	10	0	0
ptak 'bird'	62	19	23	1	1
lampa 'lamp'	54	4	4	0	0
podłoga 'floor'	49	4	8	1	0
góra 'mountain/up'	52	11	20	4	1
orzeł 'eagle'	41	10	14	1	0
owoc 'fruit'	49	15	30	3	1
mężczyzna 'male'	56	38	61	5	0
głowa 'head'	62	30	27	3	0
kapusta 'cabbage'	43	6	11	2	0
choroba 'illness'	65	22	30	1	0
łóżko 'bed'	42	7	7	1	1
ręka 'hand'	46	12	10	1	1
wódka 'vodka'	52	10	10	0	0
mięso 'meat'	71	28	36	1	1
dziecko 'child'	64	26	27	5	0
chleb 'bread'	59	15	17	0	0
praca 'work'	70	44	56	6	3
krzesło 'chair'	41	3	7	2	0
księżyc 'moon'	45	15	18	0	0
chłopiec 'boy'	44	13	22	0	1
jedzenie 'food'	61	13	22	2	0
woda 'water'	79	43	68	8	4
miasto 'city'	61	31	37	3	2
sól 'salt'	40	10	17	2	0
chata 'cottage'	39	8	10	0	0
radość 'joy'	43	26	25	0	9
król 'king'	44	13	15	0	1
pająk 'spider'	48	4	5	1	0
dziewczyna 'girl'	51	7	9	0	0
światło 'light'	42	11	12	2	3
dom 'house/home'	48	32	35	1	0
pamięć 'memory'	64	32	22	0	3
ocean 'ocean'	37	16	17	1	0
baranina 'mutton'	40	4	5	0	0
obawa 'anxiety'	27	6	10	1	0
doktor 'doctor'	48	13	24	0	0
rzeka 'river'	50	19	20	0	2
motyl 'butterfly'	53	8	12	1	0
okno 'window'	48	16	25	3	0

Only a fraction of those associations present in the network appears in the text collection. This observation may be helpful if one uses the network as a norm to evaluate the results of pure statistical algorithms which operate solely on a text collection, e.g. (Rapp 2002), (Wettler et al. 2005), (Wandmacher et al. 2008), (Gatkowska et al. 2013), (Uhr et al. 2013).

The varying size of the sub-graph and data presented in Table 3 may provide some cognitive doubt regarding purely statistical methods employed to analyze the network in itself, e.g. (Steyvers, Tennenbaum 2005).

Finally, one cannot exclude the possibility that all these results will attract the attention of researchers who are trying to present a text as a graph (e.g. (Wu et al. 2011), (Aggarwal, Zhao 2013)), solely on the basis of a text collection.

Bibliography

Aggarwal, C. C. and Zhao, P. (2013) Towards graphical models for text processing. *Knowledge and Information Systems*, 36(1):1–21.

Deo, N. (1974) *Graph Theory with Applications to Engineering and Computer Science*. Prentice-Hall.

Gatkowska, I. (2014) Word associations as a linguistic data. In: Chruszczewski, P., Rickford, J., Buczek, K., Knapik, A. and Mianowski J. (eds.), *Languages in Contact*, Wrocław, Poland, 2012, Vol.1, pp. 79–92.

Gatkowska, I., Korzycki, M., and Lubaszewski, W. (2013) Can human association norm evaluate latent semantic analysis. In: Sharp, B. Zock, M. (eds.), *Proceedings of the 10th International Workshop on Natural Language Processing and Cognitive Science*, Marseille, France, October 15th – 16th 2013, pp. 92–104.

Haręza, M. (2014) *Automatic Text Classification with Use of Empirical Association Network*, AGH, University of Science and Technology, Poland.

Kent, G. and Rosanoff, A. J. (1910) A study of association in insanity. *American Journal of Insanity*, 67(37–96):317–390.

Kiss, G. R., Armstrong, C., Milroy, R., and Piper, J. (1973) An associative thesaurus of english and its computer analysis. In: Aitken, A. J., Bailey, R. W. and Hamilton-Smith, N. (eds.), *The Computer and Literary Studies*, pp. 153–165. University Press, Edinburgh.

Korzycki, M. (2012) A dictionary based stemming mechanism for polish. In: *Proceedings of the 9th International Workshop on Natural Language Processing and Cognitive Science*, Wrocław, Poland, 28th June – 1st July 2012, pp. 143–150.

Przepiórkowski, A., Baño, M., Górski, R., Lewandowska-Tomaszczyk, B., Łaziński, M., and Pęzik, P. (2011) National corpus of polish. In: *Proceedings of the 5th Language and Technology Conference*, Poznań, Poland, 2011, pp. 259–263.

Rapp, R. (2002) The computation of word associations: comparing syntagmatic and paradigmatic approaches. In: *Proceedings of the 19th international conference on Computational linguistics*, Taipei, 2002.

Schulte im Walde, S. and Melinger, A. (2008) An in-depth look into the co-occurrence distribution of semantic associations. *Italian Journal of Linguistics, Special Issue on From Context to Meaning: Distributional Models of the Lexicon in Linguistics and Cognitive Science*, pp. 89–128.

Steyvers, M. and Tenenbaum, J. B. (2005) The large-scale structure of semantic networks: Statistical analyses and a model of semantic growth. *Cognitive Science*, 29(1):41–78.

Uhr, P., Klahold, A., and Fathi, M. (2013) Imitation of the human ability of word association. *International Journal of Soft Computing and Software Engineering (JSCSE)*, 3(3):248–254.

Wandmacher, T., Ovchinnikova, E., and Alexandrov, T. (2008) Does latent semantic analysis reflect human associations. *European Summer School in Logic, Language and Information (ESSLLI'08)*, Hamburg, Germany, 4–15 August 2008.

Wettler, M., Rapp, R., and Sedlmeier, P. (2005) Free word associations correspond to contiguities between words in texts. *Journal of Quantitative Linguistics*, 12(2–3):111–122.

Wu, J., Xuan, Z., and Pan, D. (2011) Enhancing text representation for classification tasks with semantic graph structures. *International Journal of Innovative Computing, Information and Control (ICIC)*, 7(5):2689–2698.

Mohamed Amine Boukhaled and Jean-Gabriel Ganascia

Using Function Words for Authorship Attribution: Bag-Of-Words vs. Sequential Rules

Abstract: Authorship attribution is the task of identifying the author of a given document. Various style markers have been proposed in the literature to deal with the authorship attribution task. Frequencies of function words have been shown to be very reliable and effective for this task. However, despite the fact that they are state-of-the-art, they basically rely on the invalid *bag-of-words* assumption, which stipulates that text is a set of independent words. In this contribution, we present a comparative study on using two different types of style marker based on function words for authorship attribution. We compare the effectiveness of using sequential rules of function words as style marker that do not rely on the *bag-of-words* assumption to that of the frequency of function words which does. Our results show that the frequencies of function words outperform the sequential rules.

1 Introduction

Authorship attribution is the task of identifying the author of a given doc-ument. The authorship attribution problem can typically be formulated as follows: given a set of candidate authors for whom samples of written text are available, the task is to assign a text of unknown authorship to one of these candidate authors (Stamatatos, 2009).

This problem has been addressed mainly as a problem of multi-class dis-crimination, or as a text categorization task (Sebastiani, 2002). Text categori-zation is a useful way to organize large document collection. Authorship at-tribution, as subtask of text categorization, assumes that the categorization scheme is based on the authorial information extracted from the documents. Authorship attribution is a relatively old research field. A first scientific ap-proach to the problem was proposed in the late 19th century, in the work of Mendenhall in 1887, who studied the authorship of texts attributed to Bacon, Marlowe and Shakespeare. More

Mohamed Amine Boukhaled, Jean-Gabriel Ganascia: LIP6 (Laboratoire d'Informatique de Paris 6), Université Pierre et Marie Curie and CNRS (UMR7606), ACASA Team, 4, place Jussieu, 75252-PARIS Cedex 05 (France), e-mail: {mohamed.boukhaled, jean-gabriel.ganascia}@lip6.fr

recently, the problem of authorship attribution gained greater importance due to new applications in forensic analysis, humanities scholarship (Stamatatos, 2009). Current authorship attribution methods have two key steps: (1) an indexing step based on style markers is performed on the text using some natural language processing techniques such as such as tagging, parsing, and morphological analysis; then (2) an identification step is applied using the indexed markers to determine the most likely authorship. An optional features selection step can be employed between these two key steps to determine the most relevant markers. This selection step is done by performing some statistical measures of relevance such us mutual information or Chi-square testing.

The identification step involves using methods that fall mainly into two categories: the first category includes methods that are based on statistical analysis, such as principle component analysis (Burrows, 2002) or linear discriminant analysis (Stamatatos et al., 2001); the second category includes machine learning techniques, such as simple Markov Chain (Khmelev and Tweedie, 2001), Bayesian networks, support vector machines (SVMs) (Koppel and Schler, 2004) , (Diederich et al., 2003)) and neural networks (Ramyaa and Rasheed, 2004). SVMs, which have been used successfully in text categorization and in other classification tasks, have been shown to be the most effective attribution method (Diederich et al., 2003). This is due to the fact that SVMs are less sensitive to irrelevant features in terms of degradation in accuracy, and permit one to handle high dimensional data instances more efficiently.

To achieve high authorship attribution accuracy, one should use features that are most likely to be independent from the topic of the text. Many style markers have been used for this task from early works based on features such as sentence length and vocabulary richness (Yule, 1944) to more recent and relevant works based on function words (Holmes et al., 2001), (Zhao and Zobel, 2005)), punctuation marks (Baayen et al., 2002), part-of-speech (POS) tags (Kukushkina et al., 2001), parse trees (Gamon, 2004) and character-based features (Kešelj et al., 2003).

There is an agreement among different researchers that function words are the most reliable indicator of authorship. There are two main reasons for using function words in lieu of other markers. First, because of their high frequency in a written text, function words are very difficult to consciously control, which minimizes the risk of false attribution. The second is that function words, unlike content words, are more independent from the topic or the genre of the text, so one should not expect to find great differences of frequencies across different texts written by the same authors on different topics (Chung and Pennebaker, 2007).

However, despite the fact that function word-based markers are state-of-the-art, they are basically relying on the *bag-of-words* assumption, which stipulates that text is a set of independent words. This approach completely ignores the fact that there is a syntactic structure and latent sequential information in the text. De Roeck (2004) has shown that frequent words, including function words, do not distribute homogeneously over a text. This provides evidence of the fact that the *bag-of-words* assumption is invalid. In fact, critiques have been made in the field of authorship attribution charging that many works are based on invalid assumptions (Rudman, 1997) and that researchers are focusing on attribution techniques rather than coming up with new style markers that are more precise and based on less strong assumptions.

In an effort to develop more complex yet computationally feasible stylistic features that are more linguistically motivated, (Hoover, 2003) pointed out that exploiting the sequential information existing in the text could be a promising line of work. He proved that frequent word sequences and collocations can be used with high reliability for stylistic attribution.

In this contribution, we present a comparative study on using two different types of style marker based on function words for authorship attribution. Our aim is to compare the effectiveness of using a style marker that do not rely on the *bag-of-words* assumption to that of the frequency of function words which does. In this study, we used sequential rule of function words as style marker relying on the sequential information contained in structure of the text. We first give an overview of the sequential rule mining problem in section 2 and then describe our experimental setup in section 3. Finally, the results of the comparative study are presented in section 4.

2 Sequential Rule Extraction

Sequential data mining is a data mining subdomain introduced by (Agrawal et al., 1993) which is concerned with finding interesting characteristics and patterns in sequential databases. Sequential rule mining is one of the most important sequential data mining techniques used to extract rules describing a set of sequences. In what follows, for the sake of clarity, we will limit our definitions and annotations to those necessary to understand our experiment.

Considering a set of literals called items, denoted by $I = \{i_1, \ldots, i_n\}$ an item set is a set of items $X \subseteq I$. A sequence S (single-item sequence) is an ordered list of items, denoted by $S = \langle i_1 \ldots i_n \rangle$ where $i_1 \ldots i_n$ are items.

A sequence database *SDB* is a set of tuples (*idS*), where *id* is the sequence identifier and *S* a sequence. Interesting characteristics can be extracted from such databases using sequential rules and pattern mining

Tab. 1: Sequence database SDB.

Sequence ID	Sequence
1	⟨ a, b, d, e ⟩
2	⟨ b, c, e ⟩
3	⟨ a, b, d, e ⟩

A sequential rule $R : X \Rightarrow Y$ is defined as a relationship between two item sets X and Y such that $X \cap Y = \emptyset$. This rule can be interpreted as follow: if the item set X occurs in a sequence, the item set Y will occur afterward in the same sequence. Several algorithms have been developed to efficiently extract this type of rule, such as (Fournier-Viger and Tseng, 2011). For example, if we run this algorithm on the *SDB* containing the three sequences presented in Table 1, we will get as a result sequential rules, such us $a \Rightarrow d, e$ with support equal to 2, which means that this rule is respected by two sequences in the *SDB* (i.e., there exist two sequences of the *SDB* where we find the item a, we also find d and e afterward in the same sequence).

In our study, the text is first segmented into a set of sentences, and then each sentence is mapped into a sequence of function words appearing with order in that sentence. For example the sentence "J'aime ma maison où j'ai grandi." will be mapped to ⟨ *je, ma, où, je* ⟩ as a sequence of French function words, and "*je* ⇒ *où*"; "*ma* ⇒ *où, je*" are examples of sequential rules respected by this sequence. The whole text will produce a sequential database. The rules extracted in our study represent the cadence authors follow when using function words in their writing. This gives us more explanatory properties about the syntactic writing style of a given author than just what frequencies of function words can do.

2.1 Classification Scheme

In the current approach, each text was segmented into a set of sentences based on splitting done using the punctuation marks of the set {'.', '!', '?', ':', '...'}, then function words were extracted from each sentence to construct a sequence. The algorithm described in (Fournier-Viger and Tseng, 2011) was then used to extract sequential rules of function words sequences from each text. Each text is then represented as a vector V_K of supports of rules, such that $V_K = \{r_1 r_2, \ldots, r_K\}$ is the ordered set by decreasing normalized frequency of occurrence of the top K rules in terms of support in the training set. Each text is also represented by a vector of normalized frequencies of occurrence of function words. The normalization of the vector of frequency representing a given text was done by the size of the text.

Given the classification scheme described above, we used SVMs classifier to derive a discriminative linear model from our data. To get a reasonable estimation of the expected generalization performance, we used common measures: precision (P), recall (R), and F_1 score based on a 5-fold cross–validation as follows:

$$P = \frac{TP}{TP + FP} \tag{1}$$

$$R = \frac{TP}{TP + FN} \tag{2}$$

$$F_1 = \frac{2RP}{R + P} \tag{3}$$

where *TP* are true positives, *TN* are true negatives, *FN* are false negatives, and *FP* are false positives.

3 Experimental Setup

3.1 Data Set

For the comparison experiment, we use texts written by: Balzac, Dumas, France, Gautier, Hugo, Maupassant, Proust, Sand, Sue and Zola. This choice was motivated by our special interest in studying the classic French literature of the 19th century, and the availability of electronic texts from these authors on the Gutenberg project website[1] and in the Gallica electronic library[2].

We collected 4 novels for each author, so that the total number of novels is 40. The next step was to divide these novels into smaller pieces of texts in order to have enough data instances to train the attribution algorithm.
Researchers working on authorship attribution on literature data have been using different dividing strategies. For example, Hoover (2003) decided to take just the first 10,000 words of each novel as a single text, while Argamon and Levitan (2005) treated each chapter of each book as a separate text. Since we are considering a sentence as a sequence unit, in our experiment we chose to slice novels by the size of the smallest one in the collection in terms of number of sentences. More information about the data set used in the experiment is presented in Table 2.

1 http://www.gutenberg.org/
2 http://gallica.bnf.fr/

Tab. 2: Statistics for the data set used in our experiment.

Author Name	# of words	# of texts
Balzac, Honoré de	548778	20
Dumas, Alexandre	320263	26
France, Anatole	218499	21
Gautier, Théophile	325849	19
Hugo, Victor	584502	39
Maupassant, Guy de	186598	20
Proust, Marcel	700748	38
Sand, George	560365	51
Sue, Eugène	1076843	60
Zola, Émile	581613	67

4 Results

Results of measuring the attribution performance for the different feature sets presented in our experiment setup are summarized in Table 3. These results show in general a better performance when using function words frequencies, which achieved a nearly perfect attribution, over features based on sequential rules for our corpus.

Our study here shows that the SVMs classifier combined with features extracted using sequential data mining techniques can achieve a high attribution performance (That is, $F_1 = 0.947$ for Top 400 FW-SR). Until certain limit, adding more rules increases the attribution performance.

Tab. 3: 5-fold cross-validation results for our data set. FW-SR refers to **S**equential **R**ules of **F**unctions **W**ords.

Feature set	P	R	F_1
Top 100 FW-SR	0.901	0.886	0.893
Top 200 FW-SR	0.942	0.933	0.937
Top 300 FW-SR	0.940	0.939	0.939
Top 400 FW-SR	0,951	0,944	0,947
Top 500 FW-SR	0,947	0,941	0,943
FW frequencies	**0.990**	**0.988**	**0.988**

But contrary to common sense, function-word-frequency features, which fall under the *bag-of-word* assumption known to be blind to sequential information, outperform features extracted using sequential rule mining technique. In fact, they achieved nearly a perfect performance. We believe that this due to the presence of some parameters affecting the attribution process. These parameters, that need to be more deeply studied, depend on the linguistic character of the text, such as the syntactic and the lexical differences between narrative and dialogue texts. Finally, these results are in line with previous works that claimed that bag-of-words-based features are very effective indicator of the stylistic character of a text that can enable more accurate text attribution (Argamon and Levitan, 2005).

5 Conclusion

In this contribution, we present a comparative study on using two different types of style marker based on function words for authorship attribution. We compared the effectiveness of using sequential rules of function words as style marker that do not rely on the *bag-of-words* assumption to that of the frequency of function words which does. To evaluate the effectiveness of these markers, we conducted an experiment on a classic French corpus. Our results show that contrary to common sense, the frequencies of function words outperformed the sequential rules. Based on the current study, we have identified several future research directions. First, we will explore the effectiveness of using probabilistic heuristics to ?nd a minimal sequential rule set that still allows good attribution performance, which can be very useful for stylistic and psycholinguistic analysis. Second, this study will be expanded to include sequential patterns (n-grams with gaps) as sequential style markers. Third, we intend to experiment with this new type of style markers for other languages and text sizes using standard corpora employed in the field at large.

Bibliography

Agrawal, R., Imieli'nski, T., and Swami, A. (1993). Mining association rules between sets of items in large databases. In *ACM SIGMOD Record* (Vol. 22, pp. 207–216).

Argamon, S., and Levitan, S. (2005). Measuring the usefulness of function words for authorship attribution. *In Proceedings of the Joint Conference of the Association for Computers and the Humanities and the Association for Literary and Linguistic Computing.*

Baayen, H., van Halteren, H., Neijt, A., and Tweedie, F. (2002). An experiment in authorship attribution. *In 6th JADT* (pp. 29–37).

Burrows, J. (2002). "Delta": A measure of stylistic difference and a guide to likely authorship. *Literary and Linguistic Computing*, 17(3), 267–287.

Chung, C., and Pennebaker, J. W. (2007). The psychological functions of function words. *Social Communication*, 343–359.
De Roeck, A., Sarkar, A., and Garthwaite, P. (2004). Frequent Term Distribution Measures for Dataset Profiling. *In LREC*.
Diederich, J., Kindermann, J., Leopold, E., and Paass, G. (2003). Authorship attribution with support vector machines. *Applied Intelligence*, 19(1–2), 109–123.
Fournier-Viger, P., and Tseng, V. S. (2011). Mining top-k sequential rules. *In Advanced Data Mining and Applications* (pp. 180–194). Springer.
Gamon, M. (2004). Linguistic correlates of style: authorship classification with deep linguistic analysis features. *In Proceedings of the 20th international conference on Computational Linguistics* (p. 611).
Holmes, D. I., Robertson, M., and Paez, R. (2001). Stephen Crane and the New-York Tribune: A case study in traditional and non-traditional authorship attribution. *Computers and the Humanities*, 35(3), 315–331.
Hoover, D. L. (2003). Frequent collocations and authorial style. *Literary and Linguistic Computing*, 18(3), 261–286.
Kešelj, V., Peng, F., Cercone, N., and Thomas, C. (2003). N-gram-based author profiles for authorship attribution. *In Proceedings of the conference pacific association for computational linguistics, PACLING* (Vol. 3, pp. 255–264).
Khmelev, D. V, and Tweedie, F. J. (2001). Using Markov Chains for Identification of Writer. *Literary and Linguistic Computing*, 16(3), 299–307.
Koppel, M., and Schler, J. (2004). Authorship verification as a one-class classification problem. *In Proceedings of the twenty-first international conference on Machine learning* (p. 62).
Kukushkina, O. V, Polikarpov, A. A., and Khmelev, D. V. (2001). Using literal and grammatical statistics for authorship attribution. *Problems of Information Transmission*, 37(2), 172–184.
Ramyaa, C. H., and Rasheed, K. (2004). Using machine learning techniques for stylometry. *In Proceedings of International Conference on Machine Learning*.
Rudman, J. (1997). The state of authorship attribution studies: Some problems and solutions. *Computers and the Humanities*, 31(4), 351–365.
Sebastiani, F. (2002). Machine learning in automated text categorization. *ACM Computing Surveys (CSUR)*, 34(1), 1–47.
Stamatatos, E. (2009). A survey of modern authorship attribution methods. *Journal of the American Society for Information Science and Technology*, 60(3), 538–556.
Stamatatos, E., Fakotakis, N., and Kokkinakis, G. (2001). Computer-based authorship attribution without lexical measures. *Computers and the Humanities*, 35(2), 193–214.
Yule, G. U. (1944). The statistical study of literary vocabulary. *CUP Archive*.
Zhao, Y., and Zobel, J. (2005). Effective and scalable authorship attribution using function words. *In Information Retrieval Technology* (pp. 174–189). Springer.

Daniela Gîfu and Radu Topor

Recognition of Discursive Verbal Politeness

Abstract: The paper describes a pilot study on the exploration of discursive verbal politeness in the public sphere. The purpose of this paper is to create a Romanian Gold Corpus, starting from the texts signed by public actors (journalists and politicians). A corpus of online articles is collected, in which discursive politeness is annotated on Verbal Phrases. Discursive politeness considers both dimensions of politeness, positive and negative. The study is meant to create a tool for automatic recognition of verbal politeness for any kind of public text. Furthermore, automatically annotated corpora of discourses could become complementary sources with the range of communication expertise required to reveal different valences of politeness in the public space.

1 Introduction

Since the second half of the last century, discursive politeness has been considered by Hill et al. (1986) "one of the constraints of human interaction". It has become an important topic on the public scene and, also, an interdisciplinary field of research, where pragmatics, sociolinguistics, discourse analysis, psychology, cognition, etc. intertwine. We mention here some relevant names: Grundy (2000), Watts (2003), Locher (2004), Spencer-Oatey (2002), Bargiela-Chiappini (2009).

The discursive forms that politeness can take in society vary enormously. This paper proposes a detection of discursive politeness method, by exclusive annotating of verbal markers, in order to implement an automatic recognizer of them. Furthermore, we have noticed that two politeness dimensions can be identified, known in literature as politeness and impoliteness. Mills (2011) describes these politeness dimensions. In this research study, discursive politeness was categorized in two classes: positive politeness and negative politeness.

The paper is structured in 5 sections. After a brief introduction about this topic, in section 2, the most important works on discursive politeness are retained.

Daniela Gîfu: "Alexandru Ioan Cuza" University, Faculty of Computer Science, 16, General Berthelot, 700483, Iaşi, Romania, e-mail: daniela.gifu@info.uaic.ro and University of Craiova, Center for Advanced Research in Applied Informatics, 13, A.I. Cuza St., 200585, Craiova, Romania
Radu Topor: "Alexandru Ioan Cuza" University, Faculty of Computer Science, 16, General Berthelot, 700483, Iaşi, Romania, e-mail: adutopor@info.uaic.ro

Section 3 describes the discursive politeness role in the public space, highlighting the two components (positive and negative). Section 4 presents a methodology in order to implement a tool and section 5 – a few conclusions and discussions for future work.

2 Background

In the '70–'80s of the last century, the polite verbal expression "begins to act as an epistemic object" Carpov (2007), which linguists want to explore, directly related to their interest in social interactions and their hypostases during the communication process. In line with this topic remain a series of authors like: Brown and Levinson (1978, 1987), Lakoff (1972), Leech (1983), Gofmann (1967), Grice (1975), Fraser (1975).

In November 1998, in Belgium, a colloquium was organized where the topic of verbal markers of politeness was discussed and published by Wauthion and Simon (2000), starting with the most debated relation between *politeness and ideology*. The concept of politeness is used as a strategy for eliminating any risk and for minimizing hostility, Kasper (1990), and it was later developed in France by Kerbrat-Orecchioni (1992). In fact, it is about pragmatics' influence principles: *the principle of cooperation*, with Grice's four conversational maxims, published in Grice (1975) – *quantity*, *quality*, *relation* and *manner* – which is the condition for the existence of communication; *the principle of politeness*, developed by Lakoff (1977) – *don't impose*, *give options* and *make your receiver feel good* –, Leech (1983) and then by Brown & Levinson (1978, 1987) to ensure the "anchoring" of discourse in the social dimension and, finally, *the principle of stylistic*, which refers to the promotion of the free manifestation of enunciation, showing how interpretation is formed or how new aspects of interpretation are revealed, Short (1995).

In these research papers, we observed that the preoccupations in terms of computational linguistics are insignificant. This is the reason why we have decided to implement a computational tool which would recognize discourse politeness which appears difficult to be examined.

In this work, we explore the applicability of computational methods to the recognition of Verbal Phrases (VP) in order to express politeness in public speech. Moreover, we investigate whether automatic classification techniques represent a viable approach to distinguish between positive and negative politeness, especially in political discourses.

3 Why Discursive Verbal Politeness?

For every politeness act, verbal or not, the primary function is to establish social harmony between human beings in their interactions. Due to these properties, politeness is a way to acquire a neutral expression which is not an insignificant manner to speak. Neutrality should not be confused with indifference or hypocrisy. Politeness – said Lakoff (1972) – is a matter of behavior (linguistic or any other type), "appropriate in special situations, in an attempt to achieve and to maintain successful social relationships with the others".

Most of the discursive ways in which the public orator tries to persuade bring positive or negative faces to the auditor. These can be attenuated by using specific strategies of positive politeness (e.g. exaggeration of sympathy for the receiver) or negative politeness specific strategies (e.g. a deferential attitude which implies reducing their own personality).

We also talk about politeness vs. impoliteness. Based on the principle of pragmatics of politeness by Brown and Levinson (1978, 1987), Culpeper (1996) provides an anatomy of impoliteness, regarded as an aggressive act of speech. Culpeper's strategies are intended to reveal the role of the attack on conversational balance. Watts (2003) not only confirms the importance of impoliteness, but also argues that it "should be the central point when designing a theory of politeness".

3.1 Discursive Politeness Dimensions

Considering the distance between interlocutors, we mention two types of discursive politeness:

- *Negative politeness*, as an expression of restraint, maintains the distances between human beings, focusing on the negative self-gratification of the participants in their conversations. It is a modality by which interlocutors interpose, being achievable especially in the social area.
- *Positive politeness*, as an expression of solidarity reflects the proximity between speakers, the transmitter manifesting one's sincere desire to receive positive self-gratification from one's receptor. In other words, the first, who has started a communication action, is all interested in cooperating, trying to attract the receptor in this interaction. It is about personal area.

For instance, a communication crisis can be tempered by appealing to negative politeness strategies (a deferential attitude by reducing one's own personality, the freedom of decision offered to the audience, the conventional indirect expression having an illocutionary force, etc.), or by preferring positive politeness strategies (using some identity markers which confirm the transmitter's affiliation to the group which he addresses to, exaggerating his sympathy for the receiver, etc.).

3.2 Categorization of the Verbal Phrase

Based on manual annotations of verbal markers – firstly, made experimentally – we noticed that only some of them can be marked with the dimension of politeness. These verbal politeness markers were classified into positive and negative depending on the value that the clause received (see Figure 1). Moreover, on the basis of auxiliary tools[1] that offer sentiment analysis services applied to NP (Noun Phrases) and VP (Verbal Phrases), we have done the automatic categorization of PVP (Politeness Verbal Phrase) into positive and negative.

Figure 1 shows a categorization of the VP in the corpus. They could include politeness or not (PVP and Non-PVP), and the PVP (Politeness Verbal Phrases) could be positive and negative (PPVP and NPVP).

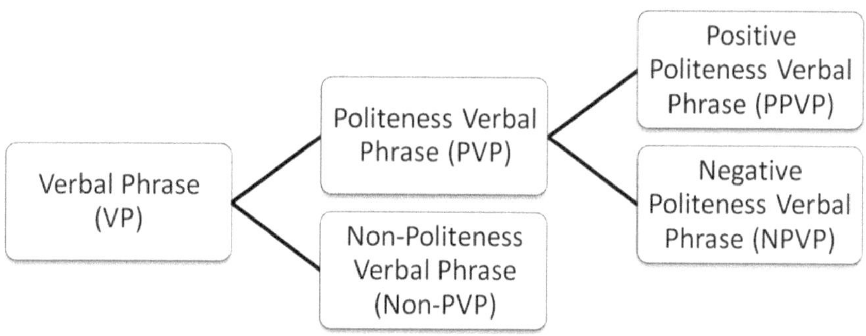

Fig. 1: A categorization of the Verbal Phrases.

4 Methodology

A corpus with online public texts (journalistic, political texts) has been built, preprocessed and annotated manually at the level of verbal structures which could define some patterns for the automatic recognition of verbal discursive politeness.

1 http://nlptools.atrilla.net/web/api.php

In addition, the corpus was automatically annotated; afterwards those two types of annotations were statistically evaluated.

In order to efficiently apply automatic tagging of Politeness Verbal Phrases within a given text, the algorithm will focus on the identification of syntactic structures containing three pylon components:

- A value construct, which will denote the positive or negative connotation of the emerging Polite Verbal Phrase.
- A subject entity, which will be the target of the PVP's value;
- A verbal phrase has a verb as its head. This verb is linking the referred subject (entity) to another element (adjective, adverb, etc.), each of them could be identified with a value. The verbal phrase can also stand as the PVP's value.

4.1 Pre-Processing

The following steps of the proposed work methodology have been applied:

- Text segmentation at sentence / clause level
  The input text has been split into sentences, which in turn has been segmented into clauses. The latter will form the basis for the focus units upon which further PVP tagging steps shall be applied. The output is further tokenized, preparing the data for further processing.

Given the following sample input text:
Ponta prelungeşte jenant criza din USL. – in Romanian (Ro)[2]
The corresponding subservient output data is:

```
<CLAUSE ID="c0" TYPE="SENTENCE">
  <W>Ponta</W>
  <W>prelungeşte</W>
  <W>jenant</W>
  <W>criza</W>
  <W>din</W>
  <W>USL</W>
</CLAUSE>
```

- Named Entity Recognition (NER)
  The emerging clauses from the previous step shall each be used as an input for a NER tool, in order to map all Name Entities within a clause, then to associate

2 Ponta embarrassingly extends the crisis from USL. – in English (En)

them to their corresponding metadata tag. This is an essential step towards the automatic identification of the aforementioned subject entity for a potential Polite Verbal Phrase, as they can hint the fact that the adjacent verbal phrase might be sentimentally charged and be part of a politeness construct.

Given the previous sample, this step will produce the following identification:

```
<NE TYPE="PERSON">
  <W>Ponta</W>
</NE>
[...]
<NE TYPE="ORGANIZATION">
  <W>USL</W>
</NE>
```

– Part-of-speech tagging
  The output from step *1)* is also ran through a POS tagging tool implemented by Simionescu (2012), as pre-processing for the following step. This operation also ensures the correct identification of the three key constructs of the PVP structures later on.

For the previously used example, this produces additional tagging of the following form:

```
[...]
  <W LEMMA="prelungi" POS="VERB" Number="singular" Person="third"
   Tense="present" TYPE="predicative">prelungeşte</W>
  <W LEMMA="jenant" POS="ADVERB" Number="singular"
   Gender="masculine">jenant</W>
  <W LEMMA="criza" POS="NOUN" Number="singular"
   Definiteness="yes">criza</W>
```

– Noun Phrase chunking
  Based on the previous Part-of-speech tagging, the NP segments are identified and annotated in their corresponding clauses. Following this process, the NP can be analyzed in order to assert if they can be assigned as the value construct for their corresponding clause, which may be associated with a positive or a negative connotation, in which case the clause can be proposed to contain a Polite Verbal Phrase. The association is done by applying sentiment analysis based on the pre-annotated corpus of PVP-annotated texts.

An aggregated output for the example after the current phase will be:

```
<CLAUSE ID="c0" TYPE="SENTENCE">
  <NP TYPE="PERSON">
    <W>Ponta</W>
  </NP>
    <W LEMMA="prelungi"POS="VERB"Number="singular"Person="third"
     Tense="present" TYPE="predicative">prelungeşte</W>
    <W LEMMA="jenant" POS="ADVERB" Number="singular"
     Gender="masculine">jenant</W>
  <NP>
    <W LEMMA="criza" POS="NOUN" Number="singular"
     Definiteness="yes">criza</W>
    <W LEMMA="din" POS="ADPOSITION">din</W>
    <NP TYPE="ORGANIZATION">
      <W>USL</W>
    </NP>
  </NP>
</CLAUSE>
```

- Verbal Phrase chunking
  A final automated operation of Verbal Phrase chunking was implemented as part of the automated preprocessing suite. This phase complements the previous tagging, and aims to facilitate further construct identification, using verbs and their corresponding verbal phrases as pivot points within the clause. The subsequently identified Polite Verbal Phrases are formed by a process of iterative "coagulation" around these VP, encompassing a nearby proposed value construct and its referred subject entity to form a candidate for a PVP.

The final pre-processing steps output correspond to the following format:

```
<CLAUSE ID="c0" TYPE="SENTENCE">
  <NP TYPE="PERSON">
    <W>Ponta</W>
  </NP>
  <VP>
    <W LEMMA="prelungi"POS="VERB"Number="singular"Person="third"
     Tense="present" TYPE="predicative">prelungeşte</W>
    <W LEMMA="jenant" POS="ADVERB" Number="singular"
     Gender="masculine">jenant</W>
  </VP>
```

```
  <NP>
    <W LEMMA="criza" POS="NOUN" Number="singular"
     Definiteness="yes">criza</W>
    <W LEMMA="din" POS="ADPOSITION">din</W>
    <NP TYPE="ORGANIZATION">
      <W>USL</W>
    </NP>
  </NP>
</CLAUSE>
```

4.2 Politeness Verbal Phrase Proposal

Having aligned both the pre-annotated corpus and the plain input text to the format specified in the previous section, the implemented software solution is now able to compile these pieces of data into a comprehensively coupled internal representation of constructs within the text, along with all their associated meta-data information. This allows flexible management of the syntactic and morphological components, in order to perform complex structural associations. The identification stage of the algorithm begins by iterating an already compiled input text, clause by clause.

As we have previously mentioned, the set of clauses which are likely to be represented by an encompassed Polite Verbal Phrase tend to contain syntactic structures which adhere to the proposed pattern of `Subject Entity` – `Verbal Phrase` – `Value`. However, it is worth mentioning that the entity may not be unambiguously morphologically represented in the focus clause, and that the value of a possible PVP might be hinted at within the constructs of the verbal phrase itself.

At this point, the algorithm first identifies the clauses containing a verbal phrase, which in its vicinity has noun phrases that are sufficiently small to have an appreciable possibility of being negatively or positively charged. These are preferred as they can be their resolution in this sense implies less ambiguity. The distance between these constructs, as well as the medial length of the noun phrases are manually calibrated following test iterations. Together with the corresponding verbal phrase, these represent the set of candidates to the goal of the clause which they pertain to.

The structure of the focus clause, the relation between its verbal phrase and its adjacent elements and the aforementioned value candidates are used to form a query profile for the clause, which is run against the pre-compiled corpus, such as a similarity score is computed for the internal representation of the clause and that of each of the PVP contained in the corpus. These scores are aggregated into a global score for the current query, encompassing tests for both structural adherence and positive / negative value tendency. If this score exceeds a calculated certainty threshold, the components of the clause which sparked the identification process are aggregated under a Politeness Verbal Phrase structure, retaining its proposed structure and components, and realigning the relations of the newly formed PVP to its parent clause.

The raw output of the module in its entirety follows the form:


```
<CLAUSE ID="c0" TYPE="SENTENCE">
   <PVP ID="pvp0" VAL="NEG">
   <NP TYPE="PERSON">
     <W>Ponta</W>
   </NP>
   <VP>
     <W LEMMA="prelungi" POS="VERB" Number="singular" Person="third"
      Tense="present" TYPE="predicative">prelungeşte</W>
     <W LEMMA="jenant" POS="ADVERB" Number="singular"
      Gender="masculine">jenant</W>
   </VP>
   <NP>
     <W LEMMA="criza" POS="NOUN" Number="singular"
      Definiteness="yes">criza</W>
     <W LEMMA="din" POS="ADPOSITION">din</W>
     <NP TYPE="ORGANIZATION">
       <W>USL</W>
       </NP>
     </NP>
   </PVP>
</CLAUSE>
```


4.3 Evaluation

Finally, a measure of evaluation is applied, consisting in the calculation of the Recall, Precision and F-measure for the automatically obtained tagged data, relative to the reference corpus and its manually annotated PVP tags.

With these values, Precision, Recall and F-measure could be computed, for both Positive and Negative Politeness VPs. The values are given in Table 2.

Tab. 1: Automatic and manual annotation results.

Total Number of Words	**6,043**
manual Politeness Verbal Phrases (mPVP)	371
automatic Politeness Verbal Phrases Phrases (aPVP)	412
manual Positive Politeness Verbal Phrases (mPPVP)	117
automatic Positive Politeness Verbal Phrases (aPPVP)	125
manual Negative Politeness Verbal Phrases (mNPVP)	254
automatic Negative Politeness Verbal Phrases (aNPVP)	287

Tab. 2: Statistical results for the detection algorithm.

	Precision	**Recall**	**F-measure**
PVP	82.70%	91.90%	87.10%
PPVP	83.20%	88.80%	85.90%
NPVP	75.20%	85.00%	79.80%

As shown in Table 2, the results of the automatic detection of polite verbal phrases, which have a positive and negative orientation, are rather high. The fact that positive politeness is scored better than negative politeness could be due to the special attention that we paid to annotating NP and VP values.

5 Conclusion and Discussion

In this present paper, we discussed the problem of Polite Verbal Phrase recognition, automatically classifying them into positive and negative.

Manual annotation created the identification frame of the most probable patterns of automatic recognition of Polite Verbal Phrase. Moreover, based on auxiliary tools which offer sentiment analysis services applied to NP and VP, we created a PVP automatic categorization into positive and negative.

During manual annotation, a number of rules were suggested, from left to right. On the one hand, the rules were applied in the implementation process and they generated a series of conflicts difficult to follow. On the other hand, this method based on rules offers more control over statistical results. The evaluation of our tool for the automatic recognition of discursive politeness consisted in comparing the manually annotated corpus with the training one having an 82.70% precision.

In the future, new VP rules will be applied, from right to left, to the Romanian discursive activity – for example – thus being a dynamic adaptation to English.

Bibliography

Bargiela-Chiappini F. and Haugh, M. (eds.), (2009) *Face, Communication and Social Interaction*, Equinox, London.

Brown, P., Levinson, S. (1978/1987) *Politeness. Some Universals in Language Usage*, Cambridge University Press, Cambridge.

Carpov, M. (2007) Politeţea şi convenţia de interpretare. *Convorbiri literare*. 141(8):17.

Culpeper, J. (1996) Towards an anatomy of impoliteness. *Journal of Pragmatics*, 25(3): 349–367.

Mills, S. (2011) Discursive approaches to politeness and impoliteness. In: Linguistic Politeness Research Group (eds.), *Discursive Approaches to politeness*, pp. 19–56, De Gruyter Mouton, Berlin and New York.

Fraser, B. (1975) The concept of politeness. Paper presented at the Fourth Annual Conference on *New Ways of Analyzing Variation in English* (NWAVE), Georgetown University.

Goffman, E. (1967) *Interaction Ritual: Essays on face-to-face behavior*. Doubleday Anchor, New York.

Grice, H.P. (1975) Logic and Conversation. In: Cole P. & Morgan J. L. (eds.), *Syntax and Semantics*, pp. 41–58, vol. III, Speech Acts., Academic, New York.

Grundy, P. (2000) *Doing Pragmatics*, Arnold, Londra, p. 164.

Hill, B., Shachiko I., Shoko I., Akiko K., and Tsunao O. (1986) Universals of linguistic politeness: Quantitative evidence from Japanese and American English. *Journal of Pragmatics* 10:349.

Kasper, G. (1990) Linguistic politeness: Current research issues. *Journal of Pragmatics*, 14: 194.

Kerbrat-Orecchioni, C. (1992) *Les interactions verbales,* tome II, A. Colin, Paris.

Lakoff, R. (1972) Language in Context. *Language*, Linguistic Society of America, 48(4):910.

Lakoff, R. (1977) What you can do with words: Politeness, pragmatics and performatives. In: Rogers, R., Wall, R. & Murphy, J. (eds.), *Proceedings of the Texas Conference on Performatives, Presuppositions and Implicatures*, Arlington, VA.: Center for Applied Linguistics, pp. 79–106.

Leech, G. (1983) *Principles of pragmatics*, Longman, London.

Locher M. (2004) *Power and Politeness in Action: Disagreements in Oral Communication*. Mouton de Gruyter, Berlin.

Simionescu, R. (2012) Romanian deep noun phrase chunking using graphical grammar studio. In: Moruz, M. A., Cristea, D., Tufiş, D., Iftene, A., Teodorescu, H. N. (eds.), *Proceedings of the 8th International Conference "Linguistic Resources And Tools For Processing Of The Romanian Language"*, Bucharest, Romania, 8–9 December 2011, 26–27 April 2012, pp. 135–143

Spencer-Oatey, H. (2002) Managing Rapport in Talk: Using Rapport Sensitive Incidents to Explore the Motivational Concerns Underlying the Management of Relations. *Journal of Pragmatics* 14:529–545.

Short, M. (1995) Understanding conversational undercurrents in The Ebony Tower by John Fowles. In Verdonk, P. and Webber, J-J (eds.), *Twentieth Century Fiction: From Text to Con-*

text, pp. 45–62, Routledge: London. Watts, R.J. (2003) *Politeness*, Cambridge University Press, Cambridge.

Wauthion, M. et A.C. Simon (éd.) (2000) Politesse & Idéologie. Rencontres de pragmatique et de rhétorique conversationnelles. In *Actes du colloque "Politesse et Idéologie"*, Louvain-la-Neuve (5–6 novembre 1998), Bibliothèque des Cahiers de l'Institut de Linguistique de Louvain 107, Peeters, Louvain-la-Neuve.

Nadine Glas and Catherine Pelachaud

Politeness versus Perceived Engagement: An Experimental Study

Abstract: We have looked at the impact of the perceived engagement level of the hearer on the speaker's perceived weight of his Face-Threatening Act, in human-human interaction. The outcome of this analysis will be applied to human-machine interaction, giving indications as to whether agents with human-like behaviour that interact with a less engaged user should employ stronger politeness strategies than when they interact with a more engaged user.

1 Introduction

For a range of applications we would like human-like agents to engage their users. We consider engagement as "the value that a participant in an interaction attributes to the goal of being together with the other participant(s) and of continuing the interaction" [14]: [12]. Numerous recent studies describe how an agent can influence user engagement by coordinating and synchronizing its behaviour with that of its user. Such behaviour includes gaze [17], gestures, postures, facial displays [7] and verbal behaviour [13]. One of the verbal aspects that can be coordinated with the user is the degree of expressed politeness [6]. En & Lan [8] indeed state that a successful implementation of politeness maxims is likely to improve human-agent engagement.

To gain more insight into the optimal coordination of politeness, we have conducted a perceptive study to verify the existence of a link between the speaker's perceived engagement level of the hearer, and the speaker's politeness strategies in human-human interaction. This provides us an indication of whether or not a human-like agent who wants to continue the interaction with its user needs to speak with more caution to someone who is less engaged than to someone who is very engaged in the ongoing interaction. Seen from the perspective of Brown & Levinson's politeness theory [3], we have tested the hypothesis that the speaker's assessment of the hearer's level of engagement has an impact on the politeness level that the speaker employs in addressing the hearer.

Nadine Glas: Institut Mines-Télécom, Télécom ParisTech, CNRS LTCI
Catherine Pelachaud: CNRS LTCI, Télécom ParisTech, 46 rue Barrault, 75013 Paris, France

2 Politeness Theory

Brown & Levinson's [3] (B&L) politeness theory is about saving the public self-image that every member wants to claim for himself, which is called this person's 'face'. The concept of 'face' consists of a negative face which is the want of every "competent adult member" that his actions be unimpeded by others; and a positive face which is the want of every member that his wants be desirable to at least some others. According to B&L some acts intrinsically threaten face, referred to as Face-Threatening Acts (FTAs). FTAs can be categorized into threats to the addressee's positive face (e.g. expressions of disapproval, criticism, disagreements) and threats to his negative face (e.g. orders, requests, suggestions). The speaker of an FTA can try to minimize the face-threat by employing a set of strategies: 1) without redressive action, baldly; 2) positive politeness; 3) negative politeness; 4) off record; 5) don't do the FTA. Roughly, the more dangerous the particular FTA x is, in the speaker's assessment, the more he will tend to choose the higher numbered strategy. W_x, the numerical value that measures the weightiness, i.e. danger, of the FTA x is calculated by: $W_x = D(S,H) + P(H,S) + R_x$ where $D(S,H)$ is the social distance between the speaker and the hearer, $P(H,S)$ the power that the hearer has over the speaker, and R_x the degree to which the FTA x is rated an imposition in that culture. The distance and power variables are intended as very general pan-cultural social dimensions.

3 Adding to the FTA-Weight Formula

In our view, besides a very general pan-cultural distance between participants in an interaction the level of engagement can be seen as a measure for distance as well. Considering our definition of engagement, a low level of engagement implies a temporally small value to *continue the interaction and be together with the other interaction participant(s)* and vice versa. This distance may be comparable with B&L's [3] distance variable, only this time it has a more temporal and dynamic nature. We thus formulate our hypothesis as: $W_x = D(S,H) + P(H,S) + R_x - Eng(H)$ where $Eng(H)$ is the speaker's perceived engagement level of the hearer. Related research includes André et al. [1] who modelled an agent that takes into account the perceived emotions of the user in adapting its politeness strategy; De Jong et al. [6] who described a model for the alignment of formality and politeness in a virtual guide; and Mayer et al. [11] who evaluated the perception of politeness in computer based tutors.

4 Method

From B&L's theory it is apparent that a straightforward way to infer the perceived threat of an FTA is by looking at the politeness strategy that is employed to formulate it. We thus created two conditions (interactions) in which the employed politeness strategies could be compared; one interaction in which a participant appears (highly) engaged (expressing the desire to continue the interaction and be together with the other interaction participant) and another in which he appears less engaged (expressing much less desire to continue the interaction and be together with the other). We have modelled such conditions for three different FTAs which were chosen according to the context of this research: Building a conversational virtual character that represents a visitor in a French museum. We looked at the FTAs: disagreement (in the preference for a painting), suggestion (to have a look at some other object) and request (for advice about what to see next). To ensure that the interaction participant demonstrated the desired levels of engagement but all other variables of the interaction were kept as constant as possible, we scripted the written interactions. Human judgements on the appropriate politeness strategies had to come from third party observers. The design of our experiment consisted of two steps: 1) The design of a collection of politeness strategies among which human judges could choose the most appropriate (Sect. 4.1 and 4.2); and 2) the design of the two different conditions (scenarios) in which the strategies needed to be chosen (Sect. 4.3).

4.1 Politeness Strategies

Following B&L's [3] tactics of formulating politeness strategies and taking inspiration from [6], we constructed a maximal set of French formulations for each FTA (as in Table 1 for the FTA 'disagreement'). The pronoun to address the hearer was kept constant as the less formal version *tu*. We did not design sentences with a mixture of strategies as this is a delicate matter that may even cause "painful jerks" instead of an accumulating politeness [3]. While theoretically the sentences can be ranked according to their potential of minimizing the FTA's risk in the way B&L proposed (Table 1, Column 1), in practice B&L's proposed hierarchy is not always entirely respected [1, 6]. To deal with this issue we first validated the sentences as described in the two paragraphs below.

Strategy Validation: Method. To validate the perceived weights of the politeness strategies that were to be used in our final experiment we performed a question-

Tab. 1: Sentences constructed to do the FTA 'disagreement', using different politeness strategies. Column 4 shows the results of the validation procedure, listing the average scores of overall politeness (q2 and q5). Column 5 indicates the sentences that are selected for the final experiment as well as their ranking within this experiment.

Politeness Strategy	Tactic	Sentence	Average politeness (Q2,5)	Rank nr.
Bald on record	Bald on record	**1.** Moi je n'aime pas cette peinture. *I don't like this painting.*	3.73 2.82	1
Positive politeness	Attend to the hearer's interests, wants, needs	**2.** Je comprends pourquoi tu aimes cette peinture, mais moi je ne l'aime pas. *I understand why you like this painting but I don't like it.*	5.73 5.36	–
	Be optimistic	**3.** Je pense que ça ne te dérange pas si je te dis que je n'aime pas cette peinture. *I think you won't mind if I tell you I don't like this painting.*	4.45 4.45	–
Negative politeness	Give reasons	**4.** Je n'aime pas cette peinture car l'art abstrait n'est pas pour moi. *I don't like this painting because abract art is not for me.*	4.45 3.82	–
	Nominalize	**5.** Le fait est que je n'aime pas cette peinture. *The thing is that I do not like this painting.*	4.45 3.45	2
	Give deference, humble oneself	**6.** Je ne suis pas expert en art abstrait mais je n'aime pas cette peinture. *I am not an expert of abstract art but I do not like this painting.*	5.09 4.64	–
	Be conventionally indirect	**7.** Je crois que je n'aime pas cette peinture. *I think I do not like this painting.*	4.82 5.27	3
	Minimize imposition	**8.** Je ne suis pas sûre d'aimer cette peinture tant que ça. *I am not sure I like this painting that much.*	5.73 5.73	5
	Apologize	**9.** Je suis désolée mais je n'aime pas cette peinture. *I am sorry but I do not like this painting.*	5.73 5.09	–
Off Record	Over generalize	**10.** En général l'art abstrait n'est pas pour moi. *In general, abstract art is not for me.*	4.91 4.64	–
	Understate	**11.** Cette peinture n'est pas laide. *This painting is not ugly.*	4.82 4.73	–
	Contradictions	**12.** J'aime bien et je n'aime pas à la fois. *I like and I don't like it.*	4.91 4.91	4

naire based survey. This questionnaire consisted of three parts, corresponding to the three different FTAs. Every part first introduced the context in which the sentences were supposed to be uttered (*Two young women in a museum meet thanks to a mutual friend.*), as well as the intention of the speaker (expressing a contradictory opinion, making a suggestion to have a look at a painting, or asking for advice about what to see next). In this way we avoided any speculation regarding the 'distance', 'power' and 'ranking' variables from B&L's formula. Below each introduction we listed all corresponding sentences. For the FTA 'disagreement' this means all sentences from Table 1, Column 3.

Every sentence in the questionnaire was followed by a set of 6 questions regarding their plausibility (yes or no, q1) and politeness level. For the latter we asked for rankings on a 7 point scale regarding respectively q2) the sentences' politeness level directly [6]; q3) the degree to which the speaker allowed the hearer to make his own decision [11] (negative politeness); q4) the degree to which the speaker wanted to work with and appreciated the hearer [11] (positive politeness); and q5) the degree to which the speaker spared the needs or face of the hearer. [1] We made sure that the order in which the FTAs were listed fluctuated between participants and that each subject was presented with a unique order of sentences. Also repetition of sentence sequences and positions was avoided.

Strategy Validation: Results. 13 subjects participated in the validation study: 5 female, all native speakers of French, aged 23–40. ANOVAs for repeated measures conducted on the ranking of overall politeness (resp. q2 and q5) showed that for each FTA the sentences differ significantly from each other (disagreement $F = 4.07$, $F = 5.84$; suggestion $F = 7.19$, $F = 8.01$; request $F = 14.38$, $F = 13.32$; $p < 0.01$). Our results confirm other studies [1, 6] that B&L's ranking of politeness according to their strategies is not completely respected. Similar to observations by De Jong et al. [6] indirect (off-record) strategies were rated much less polite than expected. The answers to validation questions q3 and q4 did not play a role in the estimation of every sentence's overall politeness level but may serve in the future to provide more insight into the outcomes of the final experiment.

4.2 Strategy Selection

From the sentences tested in the validation experiment a subset was selected for the final experiment. In this process we first eliminated those sentences that were judged more than once as implausible in the sense of an infrequent oral formula-

1 Literally *ménager la susceptibilité.*

tion. For the remaining ones we looked primarily at the mean and standard deviation of their perceived politeness level (q2). In this way we selected one sentence for every observed level of politeness. As we wanted to verify if different hearer engagement levels require different levels of politeness strategies, for our final experiment we did not require exact estimations of every strategy's politeness. Instead, the sentences were only attributed a politeness ranking relative to one another (as in Table 1, Column 5). The performed validation procedure could not take into account B&L's heaviest risk minimizing strategy of not doing the FTA at all. However, considering the nature of this strategy as complete avoidance of the FTA, we indeed assumed that there is no strategy heavier than this one and added it as such to the strategies that were used in the final experiment. E.g. for the FTA 'disagreement' (Table 1) this means that the strategy *not expressing his opinion about the painting* was added with ranking number 6.

4.3 Engagement Conditions & Questionnaire

The validation procedure had provided the politeness strategies which were to be tested for appropriateness in two different conditions. For every FTA we then constructed the two versions of dialogue that provided the contexts for these strategies. The contributions of the speaker of the FTA, say Person A, stayed constant over both conditions, while the contributions of the other, say Person B, differed considerably in form in order to communicate a high or low level of engagement. Table 2 shows the two conditions (contexts) to place the FTA 'disagreement'. For the context in which Person B was minimally engaged her utterances were kept as brief and uninterested as possible. Her engagement level was just high enough to participate in the interaction so far. In the interactions where Person B was to demonstrate a high level of engagement we used cues that have been linked to engagement in former studies and which can be expressed in written text: We made Person's B reactions longer so as to extend the interaction time [2], we added more feedback [10], added expressions of emotion [12] and of liking their interaction partner [2], and showed interest in Person A [12]. We presented human judges with a questionnaire showing one condition (interaction) for every FTA. Each condition was introduced by the same context description that was used for the validation procedure: *two young women meet in a museum through a mutual friend and start to talk.* Below each context description followed the conversations after which the participants were asked to recommend one of the sentences (each representing a politeness strategy) to Person A, under the instruction that this participant wanted to place the FTA (communicate his disagreeing opinion, do a suggestion, or ask for advice) but also wanted to absolutely continue the conversation with his interaction partner, Person B. Concretely, for the FTA "disagreement" this question looks as follows (translated):

> Q1. *Pauline does not like the Guernica (a painting by Picasso). But she absolutely wants to continue the conversation with Charlotte. Which of the following options would you advice Pauline at this moment in the conversation?*
> 1. *Alright. I don't like this painting.*
> 2. *Alright. The thing is that I do not like this painting.*
> 3. *Alright. I think I do not like this painting.*
> 4. *Alright. I like it and I don't like it.*
> 5. *Alright. I am not sure I like this painting that much.*
> 6. *Alright.*

Sentence number 6 represents the politeness strategy of not doing the FTA at all. Following Q1 we verified whether or not we had actually successfully communicated the difference between the engagement levels of person B by asking the observers to indicate on a scale from 1 to 7: Q2) the value that B attributed to the goal of being together with A; and Q3) the value that B attributed to the goal of continuing the interaction [14]: [12]. These are the direct measurements for engagement. We also asked respectively for Person B's level of involvement, rapport and interest: Q4) to what extent the interaction seemed engaging for Person B: [17]; Q5) if Person A and Person B wanted to become friends [16]); and Q6 if Person B seemed interested in the interaction.

5 Results

200 subjects participated in our final questionnaire: 68.5% female, 100% native speakers of French, aged 16–75. For every FTA observers perceived the hearer's engagement, involvement, rapport and interest levels significantly higher in the engaged condition than in the less engaged condition (t-tests $p < 0.01$). Mann-Whitney U tests on each FTA show no significant differences in the distributions of recommended politeness strategy between the two conditions (Q1, Fig. 1). For an inter-participant analysis we took the data for both conditions together and compared the rank of the participants' selected politeness strategies with their evaluation scores on engagement and related concepts (Fig. 2). Kendall Tau tests show significant negative correlations ($p < 0.05$), for the FTA 'request', between the rank of the chosen politeness strategy and the level of engagement ($\tau = -0.127$, Q2; $\tau = -0.111$, Q3), involvement ($\tau = -0.110$) and interest ($\tau = -0.107$). Similarly, the FTA 'suggestion' holds a significant negative correlation regarding the perceived level of involvement ($\tau = -0.109$).

Tab. 2: The two scenarios preceding the FTA 'disagreement'. Pauline's (Person A) utterances are similar over the two scenarios. Charlotte (Person B) shows either more (+) or less (–) engagement.

Person	Eng. level	Utterance	Engagement strategy
Pauline:		Bonjour. Je suis Pauline. Ravie de te rencontrer ! *Hello. I am Pauline. Nice to meet you!*	
Charlotte :	–	Bonjour, Charlotte. *Hello, Charlotte.*	
	+	Bonjour. Je suis Charlotte. Ravie de te rencontrer aussi. *Hello. I am Charlotte. Nice to meet you too.*	Extend reaction (Bickmore et al., 2013), expression of liking the other (Bickmore et al., 2013).
Pauline:		Heureusement vous avez réussi à venir aujourd'hui. Caroline m'a dit que vous avez eu des problèmes de transport. *Luckily you were able to make it today. Caroline told me you had some transport problems.*	
Charlotte:	–	Oui. *Yes*	
(Pauline):		Qu'est-ce qui s'est passé ? *What happened?*	
(Charlotte):		Le train avait du retard. *The train was late*	
Charlotte:	+	Oui, c'est vrai ! Notre train avait une heure de retard. *Yes, that's true! Our train was one hour late.*	Extend reaction (Bickmore et al., 2013).
Pauline:		Oh c'est gênant ! Mais, aujourd'hui le musée est ouvert jusqu'à 22 heures donc il n'y a pas de problème. *Oh that's annoying! But today the museum is open until ten o'clock so no problem.*	
Charlotte:	–	Oui. Yes.	
	+	Oui, on a de la chance. *Yes, we are lucky.*	Extend reaction (Bickmore et al., 2013), expression of emotion (Peters et al., 2005).
Pauline:		C'est la première fois que tu viens ? *Is this the first time you come here?*	
Charlotte:	–	La deuxième. *Second.*	
	+	Non, c'est la deuxième fois et toi ? *No, this is my second time, and you?*	Extend reaction (Bickmore et al., 2013), show interest in the other (Peters et al., 2005).

Tab. 2: continued.

Person	Eng. level	Utterance	Engagement strategy
Pauline:		Moi c'est ma première visite. *For me it's my first visit.*	
Charlotte:	–	–	
	+	Ah oui ? J'adore ce musée : je l'ai déjà visité l'année dernière. *Really? I love this museum. I've already visited it last year.*	Add feedback (Gratch et al., 2006), expression of emotion (Peters et al., 2005).
(Pauline):		Ah c'est super ! *That's great!*	
Pauline:		Pour l'instant je n'ai vu qu'une seule partie du musée, mais j'ai beaucoup aimé. *At the moment I've only seen part of the museum, but I like it a lot.*	
Charlotte:	–	–	
	+	Ah bon ? *Really?*	Add feedback (Gratch et al., 2006).
(Pauline):		Oui ! *Yes!*	
Pauline:		Et toi tu as déjà tout vu ? *And you, have you seen everything already?*	
Charlotte:	–	Oui. *Yes.*	
	+	Oui. Moi, j'ai tout vu. *Yes. I have seen everything.*	Extend reaction (Bickmore et al., 2013).
Pauline:		Ah c'est bien. Et qu'est-ce que tu as aimé le plus ? *Ah that's nice. And what did you like most?*	
Charlotte:	–	Guernica. *Guernica.*	
	+	Guernica, la grande peinture de Picasso. *Guernica, the big painting by Picasso.*	Extend reaction (Bickmore et al., 2013).

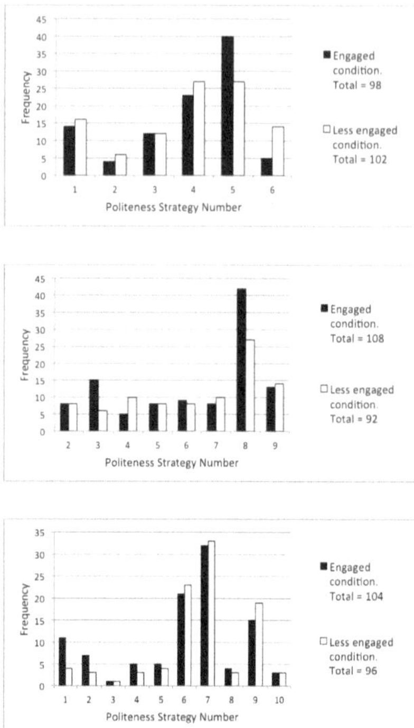

Fig. 1: Distributions of the recommended politeness strategies (Q1) for resp. the FTAs 'disagreement', 'suggestion' and 'request'.

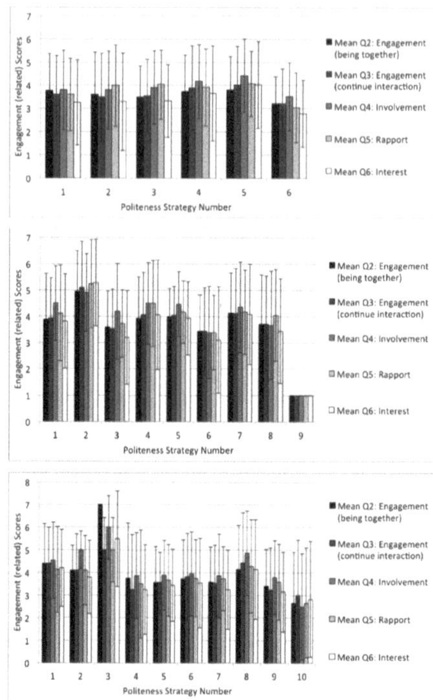

Fig. 2: Average engagement (related) scores per selected politeness strategy for resp. the FTAs 'disagreement', 'suggestion' and 'request'.

6 Discussion

The results show that we have successfully created two conditions between which the perceived engagement level of Person B is different. However, the results do not demonstrate that the recommendation of politeness strategies differs between both conditions. The lack of such a clear overall difference confirms the idea that politeness is a highly subjective phenomenon [5]. Given also that a perceived level of engagement is a highly subjective matter, we performed a finer grained, inter-participant analysis to give us more insight into the relationship between both concepts. More specifically, we compared the ranking of a participant's recommended politeness strategy with the level of engagement and related concepts he perceived in person B. It must be noted that the data points are not equally distributed as not every politeness strategy was chosen with the same frequency. Significant negative correlations were revealed in the contexts of the negative FTAs 'request' and 'suggestion'. The only FTA in our experiment that is a threat to the addressee's positive face, namely 'disagreement', does not show such correlations. A possible explanation for this is that here such a tendency interferes with a preference for alignment. This is because, a low level of engagement is conveyed by features that overlap with features that indicate positive impoliteness (e.g. "ignore or fail to attend to the addressee's interests" and "being disinterested and unconcerned" [15]). Some people prefer strong alignment settings [6] and may thus be inclined to answer positive impoliteness with less caution for the addressee's positive face as well. This tendency might interfere with an opposing preference to employ stronger politeness strategies when the addressee seems to have a lower level of engagement. This interference might explain why no significant results were found regarding a preference for alignment in [6] nor for engagement in the current experiment.

For the negative FTA 'request' we have found negative correlations between the ranking of politeness strategies and the level of engagement, involvement and interest. Specific research towards 'rapport' will be needed to explain why rapport (as in [16]) does not show a similar correlation.

The other negative FTA 'suggestion' shows only one such negative correlation, that regarding the perceived level of involvement. This may be due to the fact that the FTA can be interpreted as not really face-threatening. While B&L [3] categorize suggestions as FTAs, they may be interpreted as a remark that is placed purely in the interest of the hearer, which removes its face-threatening aspect. To a smaller extent the correlations found for the FTA 'request' might be weakened due to the fact that asking someone for his advice on what to see next implies an interest in the addressee's values and knowledge, thereby diminishing the threat.

Further it must be noted that we only considered verbal behaviour while non-verbal behaviour such as prosody, gaze and gestures can influence the way in which verbal behaviour is interpreted and reveal a range of information about the person's attitude and perceptions [12, 17]. Non-verbal expressions of feedback and mimicry for example, can play a large role in building and creating rapport [10]. Future research will have to show if third party human observers' perception will be altered in a multimodal – and so perhaps more vivid – context.

Additionally, in our current study we only integrated some aspects of engagement. Person B's contributions differ mainly in form and reaction length which might particularly influence engagement dimensions such as rapport [10] and interpersonal aspects such as warmth [9]. An emphasis on engagement aspects like paying attention [17] or showing empathy [4] may need variations in verbal and non-verbal behaviour in terms of frequency and timing. In the future we plan to verify with a multimodal approach if a focus on other dimensions of engagement will influence the results. A further analysis of the data based on the participants' gender and age is left for future research as well.

7 Conclusion

In this study we have tested if the speaker's perceived engagement level of the addressee influences the speaker's perceived weight of his FTAs. We have done this by verifying if there is a difference between the weight of the politeness strategy that a speaker would use in interaction with an engaged as compared with a less engaged person. In the creation of both conditions we have demonstrated a successful verbal behaviour model to convey a participant's engagement level. We have not found a significant overall difference between the recommendation of politeness strategies over both conditions. Politeness remains a highly subjective phenomenon with large individual differences. We did, however, find that in the context of a particular negative FTA participants who recommend weightier politeness strategies, tend to perceive a lower level of the addressee's engagement level, and vice versa. In these contexts, our hypothesis $W_x = D(S,H) + P(H,S) + R_x - Eng(H)$ seems confirmed, giving indications that an agent that wants to continue the interaction with its user needs to speak more politely to someone who is less engaged than to someone who is very engaged in the ongoing interaction. Future research will have to show to what extent these findings can be generalized.

Acknowledgments

This research was partially funded by the French DGCIS project "Avatar 1:1".

Bibliography

[1] Andre, E., Rehm, M., Minker, W., Bühler, D. 2004. *Endowing spoken language dialogue systems with emotional intelligence.* Affective Dialogue Systems. 178–187. Springer.
[2] Bickmore, T. W., Vardoulakis, L. M. P., Schulman, D. 2013. *Tinker: a relational agent museum guide.* Autonomous agents and multi-agent systems. 27(2):254–276. Springer.
[3] Brown, P., Levinson, S.C. 1987. *Politeness: Some universals in language usage.* Cambridge University Press.
[4] Castellano, G., Paiva, A., Kappas, A., Aylett, R., Hastie, H., Barendregt, W., Nabais, F., Bull, S. 2013. *Towards empathic virtual and robotic tutors.* Artificial Intelligence in Education. 733–736 Springer.
[5] Danescu-Niculescu-Mizil, C., Sudhof, M., Jurafsky, D., Leskovec, J., Potts, C. 2013. *A computational approach to politeness with application to social factors.* arXiv preprint arXiv:1306.6078.

[6] De Jong, M., Theune, M., Hofs, D. 2008. *Politeness and alignment in dialogues with a virtual guide.* Proceedings of the 7th international joint conference on Autonomous agents and multiagent systems-Volume 1., 207–214.
[7] Delaherche, E., Chetouani, M., Mahdhaoui, A., Saint-Georges, C., Viaux, S., Cohen, D. 2012. *Interpersonal synchrony: A survey of evaluation methods across disciplines.* Affective Computing, IEEE Transactions on, 3(3):349–365. IEEE.
[8] En, L. Q., Lan, S. S. 2012. *Applying politeness maxims in social robotics polite dialogue.* Human-Robot Interaction (HRI), 2012 7th ACM/IEEE International Conference on, 189–190. IEEE.
[9] Fiske, S. T., Cuddy, A. J., Glick, P. 2007. *Universal dimensions of social cognition: Warmth and competence.* Trends in cognitive sciences, 77–83. Elsevier.
[10] Gratch, J., Okhmatovskaia, A., Lamothe, F., Marsella, S., Morales, M., Van der Werf, R. J., Morency, L. P. 2006. *Virtual rapport.* Intelligent Virtual Agents, 14–27. Springer.
[11] Mayer, R. E., Johnson, W. L., Shaw, E., Sandhu, S. 2006. *Constructing computer-based tutors that are socially sensitive: Politeness in educational software.* International Journal of Human-Computer Studies. 64(1):36–42. Elsevier.
[12] Peters, C., Pelachaud, C., Bevacqua, E., Mancini, M., Poggi, I. 2005. *Engagement capabilities for ecas.* AAMAS'05 workshop Creating Bonds with ECAs.
[13] Pickering, M. J., Garrod, S. 2004. *Toward a mechanistic psychology of dialogue.* Behavioral and brain sciences, 27(2):169–190. Cambridge Univ Press.
[14] Poggi, I. 2007. *Mind, hands, face and body: a goal and belief view of multimodal communication.* Weidler.
[15] Rehm, M., Krogsager, A. 2013. *Negative affect in human robot interaction' Impoliteness in unexpected encounters with robots.* RO-MAN.,45–50. IEEE.
[16] Ringeval, F., Sonderegger, A., Sauer, J., Lalanne, D. 2013. *Introducing the RECOLA multimodal corpus of remote collaborative and affective interactions.* Automatic Face and Gesture Recognition (FG), 2013 10th IEEE International Conference and Workshops on., 1–8 IEEE.
[17] Sidner, C. L., Lee, C., Kidd, C. D., Lesh, N., Rich, C. 2005. *Explorations in engagement for humans and robots.* Artificial Intelligence, 166(1):140–164. Elsevier.

Myriam Hernández A. and José M. Gómez

Sentiment, Polarity and Function Analysis in Bibliometrics: A Review

Abstract: In this paper we proposed a survey in sentiment, polarity and function analysis of citations. This is an interesting area that has had an increased development in recent years but still has plenty of room for growth and further research. The amount of scientific information in the Web makes it necessary innovate the analysis of the influence of the work of peers and leaders in the scientific community. We present an overview of general concepts, review contributions to the solution of related problems such as context identification, function and polarity classification, in order to identify some trends and suggest possible future research directions.

1 Introduction

The number of publications in science grows exponentially each passing year. To understand the evolution of several topics, researchers and scientists require locating and accessing available contributions from among large amounts of available electronic material that can only be navigated through citations.

A citation is a text that references previous work with different purposes, such as comparing, contrasting, criticizing, agreeing, or acknowledging different sources. Citations in scientific texts are usually numerous, connect pieces of research and relate authors across time and among communities.

Citation analysis is a way of evaluating the impact of an author, a published work or a scientific media. [35] established that there are two types of research in the field of citation analysis of research papers: citation count to evaluate the impact [7, 12, 14, 22, 41] and citation content analysis [17]. The advantages of citation count are the simplicity and the experience accumulated in scientometric appli-

Myriam Hernández A.: Escuela Politécnica Nacional, Facultad de Ingeniería de Sistemas, Quito, Ecuador, e-mail: myriam.hernandez@epn.edu.ec
José M. Gómez: Universidad de Alicante, Dpto de Lenguajes y Sistemas Informáticos, Alicante, España, e-mail: jmgomez@ua.es
This research work has been partially funded by the Spanish Government and the European Commission through the project, ATTOS (TIN2012-38536-C03-03), LEGOLANG (TIN2012-31224), SAM (FP7-611312) and FIRST (FP7-287607).

cations, but many authors have pointed out its weakness. One of the limitations is that the count does not difference between the weights of high and low impact citing papers. PageRank [27] partially solved this problem with a rating algorithm. [33] proposed co-citation analysis to supplement the qualitative method with a similarity measure between works A and B, counting the number of documents that cite them.

Recently, this type researchers' impact measure has been widely criticized. Bibliometric studies [29] show that incomplete, erroneous or controversial papers are most cited. This can generate perverse incentives for new researchers who may be tempted to publish although its investigation is wrong or not yet complete because this way they will receive higher number of citations [23]. In fact, it also affects the quality of very prestigious journals such as Nature, Science or Cell because they know that accepting controversial articles is very profitable to increase citation numbers. Reviews, such as those conducted by the recently awarded Nobel Prize [30] emphasize this fact. Moreover, as claimed by [32] it is more influential the quantity of articles than their quality or than the relationship between papers with a higher number of citations and the number of citations that, in turn, they receive [40]. In this context, also could be recalled the Brazilian journals case occurred lately, that used self-references to skew the JCR index [38].

Another limitation of this method is that a citation is interpreted as an author being influenced by the work of another, without specifying the type of influence [43] which can be misleading concerning the true impact of a citation [8, 23, 26, 29, 31, 42]. To better understand the influence of a scientific work it is advisable to broaden the range of indicators to take into account factors like the author's disposition towards the reference, because, for instance, a criticized quoted work cannot have the same weight than other that is used as starting point of a research.

These problems are added to the growing importance of impact indexes for the researchers' career. Pressure to publish seems to be the cause of increased fraud in scientific literature [11]. For these reasons, it is becoming more important to correct these problems and look for more complete metrics to evaluate researchers' relevance taking into account many other "quality" factors, one of them being the intention of the researcher when citing the work of others.

Automatic analysis of subjective criteria present in a text is known as Sentiment Analysis (SA). It is a current research topic in the area of natural language processing in the field of opinion mining and its scope includes monitoring emotions in fields as diverse as marketing, political science and economics. It is proposed that SA be applied in the study of bibliographic citations, as part of citation content analysis, to detect the intention and disposition of the citing author to the cited work, and to give additional information to complement the calculation of the estimated impact of a publication to enhance its bibliometric analysis [2, 43].

Citation content analysis examines the qualitative/subjective features of bibliographic references considering the words surrounding them as a window context whose size could fluctuate from a sentence to several paragraphs. Qualitative analysis includes syntactic and semantic language relationships through speech and natural language processing and the explicit and implicit linguistic choices in the text to infer citation function and feelings of the author regarding the cited work [43]. According [36] there is a strong connection between citation function and sentiment classification.

A combination of a quantitative and qualitative/subjective analysis would give a more complete perspective of the impact of publications in the scientific community [2]. Some methods for subjective citation analysis have been proposed by different authors, but they call for more work to achieve better results in detection, extraction and handling of citations content and to characterize in a more accurate way the profile of scientists and the criticism or acceptance of their work. [6] states that when a paper is criticized, it should have a lower or a negative weight in the calculation of bibliometric measures.

In recent years, there is an explosive development in the field of sentiment analysis [21] applied to several types of text in social media, newspapers, etc. to detect sentiments from words, phrases, sentences and documents. Although having results about the polarity or the function of a citation would have important applications in bibliometric [2] and citation-based summarization [28], relative less work has been done in this field. Some general approaches of the general sentiment analysis could be applied for citation analysis but it has special features that must be considered. According [5], a citation sentiment recognizer differs from general sentiment analysis because of the singular characteristics of citations.

In this paper, we evaluate the development of subjective, sentiment and polarity citation analysis during recent years in four lines of research that are closely related: context identification, detection of implicit citations, polarity classification and purpose (function) classification. We consider proposed approaches and trace possible future work in these areas.

2 Citation Sentiment Analysis

[36] stated that the application of subjectivity, sentiment or polarity analysis is a more complex problem when applied to citation than in other domains. Citation sentiment recognizer differs from a general sentiment analyzer because of the singular characteristics of scientific text that make more difficult to detect the author intention and the motivation behind the quote. Some of these differences are:

- Authors do not always state the citation purpose explicitly. Sentiment in citations is often hidden to avoid explicit criticism especially when it cannot be quantitatively justified [15].
- Most citation sentences does not express sentiments, most of them are neutral and objective because they are just descriptive or factual.
- Negative polarity is often expressed as the result of a comparison with the author's own work and it is presented in the paper's evaluation sections.
- There is a specific sentiment related science lexicon which is primarily compound of technical terms that diverge among scientific fields [39].
- The citation context fluctuates from one clause to several paragraphs. To delimitate the scope of a citation is a problem not yet satisfactorily solved and involves lexical, semantic and discourse considerations. [5] stated that most of the work in subjective citation analysis covers only the citation sentence and stressed the necessity to cover a wider citation context to detect polarity or function of a citation in a more accurate way. All the more, [35, 36] suggested that all related mentions of the paper are not concentrated in the surrounding area of the citation and expressed the necessity of developing natural language processing methods capable of finding and linking together citations with sentences qualifying them that are not around the citation physical location.

Furthermore, in scientific literature the citing author disposition towards a cited work is often expressed in indirect ways and often with academic expressions that are particular for each knowledge field. Efforts have being made to connect sentiments to technical terms and to develop general sentiment lexicons for Science [34].

3 Citation Context Identification and Detection of Implicit Citations

Different works which have being made to identify optimal size of context windows in order to detect the sentences referring to the citation. Most citation classification approaches are not of general application but are heavily oriented for specific science domains [10]. Almost all the research in this field corresponds to the Computer Science and Language Technologies fields and some to Life Science and Biomedical Topics. The more frequently used technique is machine learning algorithms with supervised approaches. [19] implemented a supervised tool with an algorithm using co-reference chains for extracting citation with a Super Vector

Machine (SVM) classifier handling citation strings in the MUC-7[1] dataset. [35] used supervised training methods with n-grams (unigrams and bigrams) and other features as proper nouns, previous and next sentence, position, and orthographic characteristics. Two algorithms were used for the classifier: Maximum Entropy (ME), and SVM. As a result they made subjectivity analysis and classified text as citing or non-citing. They confirmed through experimentation that the use of context compound of sentences around the citation leads to a better identification of citation sentences, they used the ACL Anthology Corpus[2].

[5] worked in a supervised approach with context windows of different lengths. They developed a classifier using citations as features sets with SVM and n-grams of length 1 to 3 as well as dependency triplets as features. They obtained a citation sentiment classification using an annotated context window of 4 sentences. Their method detected implicit and explicit citations in that context window taking into account the section of the article where citations appear (abstract, introduction, results, etc.) They specified four categories of citation sentiment: polarity, manner, certainty and knowledge type.

Conditional Random Fields (CRF) is a technique used to tag, segment and extract information from documents and have been used for some authors to identify context. [3] implemented a CRFs-based citation context extraction. They developed a context extraction tool: CitContExt. This software detects six no-context type sentences: background, issues, gaps, description, current work outcome and future work; and seven context type sentences: cited work identifies gaps, cited work overcomes gaps, uses outputs from cited works, results with cited work, compare works of cited work, shortcomings in cited work, issue related cited work.

[2] addressed the problem of identifying fragments of a citing sentence that are related to a target reference, they called it reference scope. They approach context identification as a different problem because they considered a citing sentence that cites multiple papers. They used and compared three methods: word classification SVM, sequence labeling CRF and segment classification. Segment classification was executed splitting the sentence into segments by punctuation and coordination conjunctions and was the algorithm that achieved best performance. They used a corpus formed by 19,000 NLP papers from ACL Anthology Network (ANN) corpus.

[28] experimented with probabilistic inference to extract citation context identifying even non explicit citing sentences for citation-based summarization.

---

1 Linguistic Data Consortium, University of Pennsylvania, http://catalog.ldc.upenn.edu/LDC2001T02

2 Version 20080325, http://acl-arc.comp.nus.edu.sg/

They modeled sentences as a CRF to detect the context of data patterns, employed a Belief Propagation mechanism to detect likely context sentences and used SVM to classify sentences as being inside or outside context. Best results were obtained for a window that has each sentence connected to 4 sentences on each side.

[34] implemented a context extraction method with a graphical interface in the form of global and within specialty maps that show links labeled by sentiment for citation and co-citation contexts graphically as global and within-speciality maps to be analyzed in terms of sentiment and content. The corpora were processed using bag of sample cue words that denote sentiment or function and count their frequency of appearance. They used a subset of 20 papers of the citation summary data of the AAN [28].

4 Citation Purpose Classification and Citation Polarity Classification

The fields of citation purpose classification (function) and citation polarity classification are interesting research topics that are being explored in recent years. An introduction to an automated solution to this problem was first developed for [36] applied in 116 articles randomly chosen from annotated corpus from Computation and Language E-Print Archive[3]. They used a supervised technique to automatically annotate citation text using machine learning to classify 12 citation functions aggregated in four categories. This work did not include a posterior classification of unlabeled text.

[25] used a supervised probabilistic model of classification based on a dependency tree with CRFs to distinguish positive and negative polarity to a dataset formed by four corpora for sentiment detection in Japanese.

[10] used a semi-supervised automatic data annotation with an ensemble-style self-training algorithm. Classification was aspect-based with NaÃŕve Bayes as main basic technique. Citation was classified as: background, fundamental idea, technical basis and performance comparison applied over randomly selected papers from ACL ARC[4].

3 http://xxx.lanl.gov/cmp-lg
4 http://aclweb.org/anthology

[4] implemented an aspect-based algorithm in which each citation had a feature set inside a SVM. Corpus is processed using WEKA[5] resulting in polarity citation classification.

[18] applied a faceted algorithm using Moravcsik and Murugesan annotation scheme and the Stanford ME classifier. In this approach we have a four aspects scheme and context lengths of 1, 2 and 3 sentences. Conceptual vs. operational facet asks: "Is this an idea or a tool?; organic vs. perfunctory facet distinguishes those citations that form the foundations from the citing work from more superficial citations; evolutionary vs. juxtapositional facet detects the relationship between the citing and cited papers: based on vs. alternative work; confirmative vs. negational facet, captures the completeness and correctness of the cited work according to citing work.

[24] implemented a hybrid algorithm that used discourse as a tree-model and analyzed POS as regular expressions to obtain citation relations of contrast and corroboration.

[9] obtained XML from PDF formats using PDFX and then extract citations by means of XSLT. They retrieved citation functions using citation context with CiTalO[6] tool and two ontologies: CiTO2Wordnet and CiTOfunctions. Following are some of the positive, neutral and negative functions they used: agrees with, cites, cites as authority, cites a data source, cites as evidence, cites as metadata document, cites as potential solution, cites as recommended reading, cites as related, confirms, corrects, critiques, derides.

[20] used a ME-based system to train from annotated data an automatically classify functions in three categories linked to sentiment polarity: Positive (agreement or compatibility), Neutral (background), and Negative (shows weakness of cited work).

[16] implemented citation extraction algorithm through pattern matching. They detected polarity (positive, negative and neutral) with CiTalO, a software they developed using a combination of techniques of ontology learning from natural language, sentiment analysis, word sense disambiguation (with SVM), and ontology mapping. Corpus used was 18 papers published in the seventh volume of Balisage Proceedings (2011)[7], about Web markup technologies.

[37] made unsupervised boot-strapping algorithm for identifying and categorizing in two categories of concepts: Techniques and Applications.

5 http://www.cs.waikato.ac.nz/ml/weka/
6 http://wit.istc.cnr.it:8080/tools/citalo
7 http://balisage.net/Proceedings/vol7/cover.html

[13] established a faceted classification of citation links in networks using SVM with linear kernels. They used three mutually exclusive categories: functional, perfunctory, and a class for ambiguous cases.

[1] used a trained classification model and SVM classifier with Linear Kernel applied to the AAN dataset.

5 Discussion

We have seen that the most used classification tool is machine learning followed by dictionary based-approaches. Hybrid methods, such a combination of machine learning and rule-based algorithms or dictionary methods with NLP are not very implemented despite their good results in general sentiment analysis, probably because of their complexity.

In citation analysis, sentiments are represented in a binary way, as polarity, and not in discrete scales. Functions are categorized in diverse forms that can always be mapped to their polarity.

In citation context identification, performance results for precision vary from approximately 77% to 83%. In citation purpose and polarity analysis results for precision have a greatly variation with a wide range that goes from 26% to 91%. Nevertheless, the values obtained in the different methods cannot be directly compared because the execution frames and evaluation methodologies diverge substantially among the studies.

In sentiment, polarity and function citation analysis there are open problems and interesting areas for future research such as:

- Detection of size of context windows.
- Detection of all references to cited work including those outside window context using NLP techniques and discourse analysis.
- Detection of non-explicit citing sentences for sentiment citation analysis.
- Development and application of domain independent techniques.
- Development and application of hybrid algorithms to obtain better performance.
- Development of a unique framework to execute experimental comparison among different techniques: algorithms, datasets, feature selection.
- Facet-analysis to classify sentiment, function or polarity of separate aspects and functions of the same work.

6 Conclusions

In this paper we have proposed a survey in sentiment, polarity and function analysis of citations.

Although work in this specific area has increased in recent years, there are still open problems that have not been solved and they need to be investigated. There are not enough open corpus that can be worked in shared form by researchers, there is not a common work frame to facilitate achieving results that are comparable with each other in order to reach conclusions about the efficiency of different techniques. In this field it is necessary to develop conditions that allow and motivate collaborative work.

Bibliography

[1] Amjad Abu-jbara, Jefferson Ezra, and Dragomir Radev. Purpose and polarity of citation: Towards nlp-based bibliometrics. *Proceedings of NAACL-HLT*, pages 596–606, 2013.

[2] Amjad Abu-Jbara and Dragomir Radev. Reference scope identification in citing sentences. In *12 Proceedings of the 2012 Conference of the North American Chapter of the Association for Computational Linguistics: Human Language Technologies*, pages 80–90, Stroudsburg, PA, USA, June 2012. Association for Computational Linguistics.

[3] M. A. Angrosh, Stephen Cranefield, and Nigel Stanger. Conditional random field based sentence context identification: enhancing citation services for the research community. In *Proceedings of the First Australasian Web Conference*, volume 144, pages 59–68, Adelaide, Australia, January 2013. Australian Computer Society, Inc.

[4] Awais Athar. Sentiment analysis of citations using sentence structure-based features. In *11 Proceedings of the ACL 2011 Student Session*, pages 81–87, Stroudsburg, PA, USA, June 2011. Association for Computational Linguistics.

[5] Awais Athar and Simone Teufel. Context-enhanced citation sentiment detection. In *Proceedings of the 2012 Conference of the North American Chapter of the Association for Computational Linguistics: Human Language Technologies*, pages 597–601, Montreal, Canada, June 2012. Association for Computational Linguistics.

[6] Susan Bonzi. Characteristics of a Literature as Predictors of Relatedness Between Cited and Citing Works. *Journal of the American Society for Information Science*, 33(4):208–216, September 1982.

[7] Christine L Borgman and Jonathan Furner. Scholarly communication and bibliometrics. *Annual Review of Information Science and Technology*, 36(1):2–72, February 2005.

[8] Björn Brembs and Marcus Munafò. Deep Impact: Unintended consequences of journal rank. January 2013.

[9] Paolo Ciancarini, Angelo Di Iorio, Andrea Giovanni Nuzzolese, Silvio Peroni, and Fabio Vitali. Semantic Annotation of Scholarly Documents and Citations. In Matteo Baldoni, Cristina Baroglio, Guido Boella, and Roberto Micalizio, editors, *AI*IA*, volume 8249 of *Lecture Notes in Computer Science*, pages 336–347. Springer, 2013.

[10] Cailing Dong and Ulrich Schäfer. Ensemble-style Self-training on Citation Classification. In *Proceedings of 5th International Joint Conference on Natural Language Processing*, pages 623–631, Chiang Mai, 2011. Asian Federation of Natural Language Processing.

[11] Ferric C Fang, R Grant Steen, and Arturo Casadevall. Misconduct accounts for the majority of retracted scientific publications. *Proceedings of the National Academy of Sciences of the United States of America*, 109(42):17028–33, October 2012.

[12] E Garfield. Citation Analysis as a Tool in Journal Evaluation: Journals can be ranked by frequency and impact of citations for science policy studies. *Science*, 178(4060):471–479, November 1972.

[13] Eric Martin Han Xu. Using Heterogeneous Features for Scientific Citation Classification. In *Proceedings of the 13th conference of the Pacific Association for Computational Linguistics*. Pacific Association for Computational Linguistics, 2013.

[14] J E Hirsch. An index to quantify an individual's scientific research output. *Proceedings of the National Academy of Sciences of the United States of America*, 102(46):16569–72, November 2005.

[15] K. Hyland. Writing Without Conviction? Hedging in Science Research Articles. *Applied Linguistics*, 17(4):433–454, December 1996.

[16] Angelo Di Iorio, Andrea Giovanni Nuzzolese, and Silvio Peroni. Towards the Automatic Identification of the Nature of Citations. In *SePublica*, pages 63–74, 2013.

[17] C. Lee Giles Min-yen Kan Isaac G. Councill. ParsCit: An open-source CRF reference string parsing package. In *Proceedings of the Sixth International Language Resources and Evaluation*, pages 661–667, Marrakech, Morocco, 2008. European Language Resources Association.

[18] Charles Jochim and Hinrich Schütze. Towards a Generic and Flexible Citation Classifier Based on a Faceted Classification Scheme. In *Procedings of COLING'12*, pages 1343–1358, 2012.

[19] Dain Kaplan, Ryu Iida, and Takenobu Tokunaga. Automatic extraction of citation contexts for research paper summarization: a coreference-chain based approach. In *Proceedings of the 2009 Workshop on Text and Citation Analysis for Scholarly Digital Libraries*, pages 88–95, Suntec, Singapore, August 2009. Association for Computational Linguistics.

[20] Xiang Li, Yifan He, Adam Meyers, and Ralph Grishman. Towards Fine-grained Citation Function Classification. In *RANLP*, pages 402–407, 2013.

[21] B Liu and L Zhang. A survey of opinion mining and sentiment analysis. *Mining Text Data*, pages 415–463, 2012.

[22] T Luukkonen, O Persson, and G Sivertsen. Understanding Patterns of International Scientific Collaboration. *Science, Technology & Human Values*, 17(1):101–126, January 1992.

[23] Eve Marder, Helmut Kettenmann, and Sten Grillner. Impacting our young. *Proceedings of the National Academy of Sciences of the United States of America*, 107(50):21233, December 2010.

[24] Adam Meyers. Contrasting and Corroborating Citations in Journal Articles. *Proceedings of the International Conference Recent Advances in Natural Language Processing RANLP 2013*, pages 460–466, 2013.

[25] Tetsuji Nakagawa, Kentaro Inui, and Sadao Kurohashi. Dependency tree-based sentiment classification using CRFs with hidden variables. In *Proceeding HLT '10 Human Language Technologies: The 2010 Annual Conference of the North American Chapter of the Association for Computational Linguistics*, pages 786–794, Stroudsburg, PA, USA, 2010. Association for Computational Linguistics.

[26] Joshua M Nicholson and John P A Ioannidis. Research grants: Conform and be funded. *Nature*, 492(7427):34–6, December 2012.

[27] Lawrence Page, Sergey Brin, Rajeev Motwani, and Terry Winograd. The PageRank Citation Ranking: Bringing Order to the Web., November 1999.

[28] Dragomir R. Radev, Pradeep Muthukrishnan, and Vahed Qazvinian. The ACL Anthology Network corpus. In *Proceedings of the 2009 Workshop on Text and Citation Analysis for Scholarly Digital Libraries*, pages 54–61, Suntec, Singapore, August 2009. Association for Computational Linguistics.

[29] Filippo Radicchi. In science "there is no bad publicity": papers criticized in comments have high scientific impact. *Scientific reports*, 2:815, January 2012.

[30] Ian Sample. Nobel winner declares boycott of top science journals, September 2013.

[31] Michael Schreiber. A case study of the arbitrariness of the h-index and the highly-cited-publications indicator. *Journal of Informetrics*, 7(2):379–387, April 2013.

[32] Donald Siegel and Philippe Baveye. Battling the paper glut. *Science (New York, N.Y.)*, 329(5998):1466, September 2010.

[33] Henry Small. Co-citation in the scientific literature: A new measure of the relationship between two documents. *Journal of the American Society for Information Science*, 24(4):265–269, July 1973.

[34] Henry Small. Interpreting maps of science using citation context sentiments: a preliminary investigation. *Scientometrics*, 87(2):373–388, February 2011.

[35] Kazunari Sugiyama, Tarun Kumar, Min-Yen Kan, and Ramesh C Tripathi. Identifying citing sentences in research papers using supervised learning. In *2010 International Conference on Information Retrieval & Knowledge Management (CAMP)*, pages 67–72. IEEE, March 2010.

[36] Simone Teufel, Advaith Siddharthan, and Dan Tidhar. Automatic classification of citation function. pages 103–110, July 2006.

[37] Chen-Tse Tsai, Gourab Kundu, and Dan Roth. Concept-based analysis of scientific literature. In *Proceedings of the 22nd ACM international conference on Conference on information & knowledge management – CIKM '13*, pages 1733–1738, New York, New York, USA, October 2013. ACM Press.

[38] Richard Van Noorden. Brazilian citation scheme outed. *Nature*, 500(7464):510–1, August 2013.

[39] Mateja Verlic, Gregor Stiglic, Simon Kocbek, and Peter Kokol. Sentiment in Science – A Case Study of CBMS Contributions in Years 2003 to 2007. In *2008 21st IEEE International Symposium on Computer-Based Medical Systems*, pages 138–143. IEEE, June 2008.

[40] Gregory D Webster, Peter K Jonason, and Tatiana O Schember. Hot topics and popular papers in evolutionary psychology: Analyses of title words and citation counts in Evolution and Human Behavior, 1979–2008. *Evolutionary Psychology*, 7(3):348–362, 2009.

[41] Howard D White and Katherine W McCain. Visualizing a discipline: An author co-citation analysis of information science, 1972âĂŞ1995. *Journal of the American Society for Information Science*, 49(4):327–355, 1998.

[42] Neal S Young, John P A Ioannidis, and Omar Al-Ubaydli. Why current publication practices may distort science. *PLoS medicine*, 5(10):e201, October 2008.

[43] Guo Zhang, Ying Ding, and Staša Milojević. Citation content analysis (cca): A framework for syntactic and semantic analysis of citation content. November 2012.

Manfred Klenner, Susanna Tron, Michael Amsler, and Nora Hollenstein

The Detection and Analysis of Bi-Polar Phrases and Polarity Conflicts

Abstract: In fine-grained sentiment analysis one has to deal with the composition of bi-polar phrases such as e.g. *just punishment*. Moreover, the top down prediction of phrase polarity as imposed by certain verbs on their direct objects sometimes is violated by the bottom up composed phrase polarity (e.g. 'to approve war'). We introduce a fine-grained polarity lexicon built along the lines of the Appraisal Theory and we investigate the composition of bi-polar phrases – both, from a phrase internal point of view and from a verb-centered perspective. We have specified a multi-lingual polarity resource (French, English, German) and a system pipeline that carries out sentiment composition for these languages. We discuss examples with reference to each of these languages.

1 Introduction

Advanced sentiment analysis builds on sentiment composition where the polarity of words from a prior polarity lexicon are combined to give the polarity of phrases in order to incrementally approach the polarity of the whole clause. Simple approaches, where the polarity of a phrase or sentence is calculated from the ratio of positive and negatvie word level polarities and the number of negators are bound to fail in the presence of sophisiticated language, that is, language beyond product reviews – a text genre most desparately analysed in the field of sentiment analysis so far. We argue that more fine-grained distinctions are needed to analyse the polar load of texts than reference to the basic distinctions 'positive' and 'negative'. The Appraisal Theory (Martin and White (2005)) has introduced such a finer distinction: the one between judgement, emotion and attitude. A word or phrase is no longer just positive, but might be positive on a moral (humanity), emotional (love) or factual (victory) basis.

A particular problem arises if two or more words with different polarities are combined to form a phrase. What is the phrase-level polarity, then? Take *(this is)*

Manfred Klenner, Susanna Tron, Michael Amsler, Nora Hollenstein: Institute of Computational Linguistics, University of Zurich, Switzerland, e-mail: {klenner|tron|mamsler|hollenstein}@cl.uzh.ch

just punishment – is this positive, negative or should we rather just stick with the term 'bi-polar'? If we follow the distinctions made by the Appraisal Theory, another option would be to just refer to different kinds of polarity. We could say that *just punishment* is positive at the moral level (since the punishment is described as being just) but from a factual perspective it is negative (since it mentally or physically injures a person).

Another source of polarity conflicts are verbs such as 'to admire', 'to prefer' etc. They seem to impose a polarity expectation on their arguments (e.g. the direct object). So 'admire' requires a positive or a neutral noun phrase as its direct object. But what if the noun phrase was negative? Does one admire negative objects (in the broadest sense). It depends on the kind of negativity, we argue. How can we explain why *He admires his sick friend* does not produce a conflict while *He admires his deceitful friend* does, although both noun phrases are negative (*sick friend* and *deceitful friend*). One hypothesis is that the distinction between morally negative and (just) factually negative might be the cause. Also, being sick is unintended (passive) while being deceitful is not (is active).

2 Fine-Grained Polarity Lexicon

We aim at a compositional treatment of phrase- and sentence-level polarity. In order to assure high quality, we rely on a manually crafted polarity lexicon specifying the polarities of words (not word senses). Recently, fine-grained distinctions have been proposed that distinguish between various forms of positive and negative polarities, e.g. (Neviarouskaya et al. (2009)). For instance, the Appraisal Theory (Martin and White (2005)) suggests to distinguish between appreciation (*sick friend*), judgement (*deceitful friend*) and emotion (*angry friend*).

We have adopted and slightly adapted the categories of the appraisal theory. Our French, German and English polarity lexicons comprise 15,700 single-word entries (nouns, adjectives, adverbs), manually annotated for positive and negative prior polarity where each class further specifies whether a word is factually, morally or emotionally polar. We also coded whether the word involves an active part of the related actor (where applicable) and whether it is weakly or strongly polar. Our ultimate goal is to combine this resource with our verb resources (described below) in order to predict the polarity of the verb arguments and to be able to deal with conflicts arising from violated polarity expectations of the verb.

Also part of our lexicon are shifters (inverting the polarity, e.g., *a good idea* (positive) vs. *not a good idea* (negative)) and intensifiers and diminishers. Table 1 shows the details of our (in the paper focussed) French lexicon. For the English version the following statistics holds: shifters 24, intensifiers 94, diminishers 20, positive nouns 583, negative nouns 1345, positive adjectives 1097, negative adjectives 1475. Table 2 gives an overview of our labels together with same examples: The prefixes A, F and J denote appreciation (factual), affect and judgement, respectively. The German lexicon comprises 2103 negative nouns, 1249 positive nouns, 1482 positive adjectives, 1861 negative adjectives, 71 intensifiers, 15 diminishers and 14 shifters.

Tab. 1: French polarity lexicon.

word class	NEG	POS	DIM	INT	SHI	Total
Adjectives	1550	858	3	34	5	2450
Nouns	1332	508	1	10	5	1856
ALL	2917	1411	5	79	26	4438

Tab. 2: Complete list of polarity labels with examples.

Tag	Meaning	Examples
A POS	Appreciation Positive	optimisation, beautiful, productive
F POS	Affect Positive	sensitive, happiness, love
J POS	Judgment Positive	charity, fidelity, charming
A NEG	Appreciation Negative	illness, unstable, loss
F NEG	Affect Negative	hatred, mourn, afraid
J NEG	Judgment Negative	corrupted, dictator, torture
DIM	Diminisher	less, decreasing
INT	Intensifier	more, vast
SHI	Shifter	not, absence of

3 Phrasal Level Sentiment Composition

According to the principle of compositionality and along the line of other scholars (e.g. Moilanen and Pulman (2007)), after mapping polarity from the lexicon to the text, in the next step we calculate the polarity of nominal and prepositional phrases, i.e., based on the lexical marking and taking into account the syntactic analysis of a dependency parser, we conduct a composition of polarity for the phrases. In general, the polarities are propagated bottom-up to their respective heads of the NPs/PPs in composition with the other subordinates. To conduct this composition we convert the output of the dependency parser into a constraint grammar format and use the `vislcg3`-tools VISL-group (2013) which allows us to write the compositional rules in a concise manner.

Composition at the np level is straightforward, if the polarity labels do match (including the prefix). Independent of the prefix (A,F,J) is it save to induce that a positive adjective coupled with a positive noun yields a positive noun phrase (given that no shifters are around): the prefix is left unchanced, of course. So *a lucky* (A_POS) *donator* (A_POS) gives a positive np *lucky donator* (A_POS). The same is true for the other prefixed polarities (F_POS + F_POS = F_POS etc.)

The interesting cases are those noun phrases where the adjective has a different polarity from the noun. For instance, here is a couple of noun phrases taken from the (English) Gutenberg corpus where a F_NEG adjective is combined with a F_POS noun.

Tab. 3: Examples of bi-polar phrases: F_NEG-F_POS combinations.

adjective	noun	adjective	noun
nervous	emotion	disappointed	affection
angry	passion	grief	joys
furious	passions	anxious	hopes
nervous	gratitude	angry	pleasure
disappointed	hopes	disappointed	love
angry	joy	sad	astonishment
unhappy	passions	sad	pleasure

Here, the prefix stays, but the polarity needs to be fixed to either positive or negative. It seems clear from these cases that the negative polarity wins: *disappointed hope* is negative. The same is true for F_POS-F_NEG combiniation and the J prefix variants. So always NEG wins. However, an empirical analysis should confirm these hypotheses.

But what if prefixes differ, e.g. a J_POS adjective collides with a F_NEG noun? Which prefix, which polarity should we keep? In order to make the problem behind them clear, we call these cases *bi-polar* noun phrases, although we believe that in most cases a decision can be taken: whether it is positive or negative. Even the prefix might be clear. See Table 4 for some examples. We could just stipulate a composition rule but rather found it more appropriate to base such a decision on an empirical study (see section 5).

Before we turn to the other conflict scenario, where verbs and their polarity expectations are violated, we introduce the verb resource itself.

Tab. 4: Examples of bi-polar phrases: J_POS-F_NEG combinations.

adjective	noun	adjective	noun
earnest	regret	sincerely	anxious
kindly	regret	wisest	sorrow
honest	concern	heroic	angers
noble	rage	honest	shame
lively	despair	decent	sorrow

4 Polar Verbs: Effects and Expectations

In order to merge the polar information of the NPs/PPs on the sentence level one must include their combination via their governor which is normally the verb. Neviarouskaya et al. (2009) propose a system in which special rules for verb classes relying on their semantics are applied to attitude analysis on the phrase/clause-level. Reschke and Anand (2011) show that it is possible to set the evaluativity functors for verb classes to derive the contextual evaluativity, given the polarity of the arguments. Other scholars carrying out sentiment analysis on texts that bear multiple opinions toward the same target also argue that a more complex lexicon model is needed and especially a set of rules for verbs that define how the arguments of the subcategorization frame are affected – in this special case concerning the attitudes between them, see Maks and Vossen (2012).

Next to the evidence from the mentioned literature and the respective promising results, there is also a strong clue coming from error analysis concerning sentiment calculation in which verbs are treated in the same manner as the composition for polar adjectives and nouns described above. This shows up especially if one aims at a target specific (sentence-level) sentiment analysis: in a given sentence "*Attorney X accuses Bank Y of investor fraud.*" one can easily infer that *accuse* is a verb carrying a negative polarity. But in this example the direct object *Bank Y* is accused and should therefore receive a negative "effect" while the *Attorney X* – as the subject of the verb – is not negatively affected at all. Second, the PP *of investor fraud* is a modification of the accusation (giving a reason) and there is intuitively a tendency to expect a negative polarity of this PP – otherwise the accusation would be unjust (In the example given, the negative expectation matches with the composed polarity stemming from the lexically negative "fraud"). So it is clear that the grammatical function must be first determined (by a parser) in order to accurately calculate the effects and expectations that are connected to the lexical-semantic meaning of the verb.

Furthermore, the meaning of the verb (and therefore the polarity) can change according to the context (cf. "report a profit" (positive) vs. "report a loss" (negative) vs. "report an expected outcome" (neutral)). This leads to a conditional identification of the resulting verb polarity (or verbal phrase respectively) in such a manner that the polarity calculated for the head of the object triggers the polarity of the verb. In German, for instance, there are verbs that not only change their polarity in respect to syntactic frames (e.g. in reflexive form) but also in respect to the polarity of the connected arguments, too (see Table 5).

Tab. 5: Several examples for the use of the German verb "sorgen".

German	English	Polarity
für die Kinder sorgen	to take care of the kids	positive
für Probleme[neg.] sorgen	to cause problems	negative
für Frieden[pos.] sorgen	to bring peace	positive
sich sorgen	to worry	negative

We therefore encode the impact of the verbs on polarity concerning three dimensions: effects, expectations and verb polarity. While effects should be understood as the outcome instantiated through the verb, expectations can be understood as anticipated polarities induced by the verb. The verb polarity as such is the evaluation of the whole verbal phrase. To sum up: in addition to verb polarity, we introduce effects and expectations to verb frames which are determined through the syntactic pattern found (including negation), the lexical meaning concerning polarity itself and/or the conditional polarity respective to the bottom-up calculated prevalent polarities. This results at the moment in over 120 classes of verb patterns with regard to combinations of syntactic pattern, given polarities in grammatical functions, resulting effects and expectations, and verb polarity.

As an example we take the verb class *fclass_subj_neg_objа_eff_verb_neg* which refers to the syntactic pattern (subject and direct object) and at the same time indicates which effects and/or expectations are triggered (here negative effect for the direct object). If the lemma of the verb is found and the syntactic pattern is matched in the linguistic analysis, then we apply the rule and assign the impacts to the related instances. However, the boundary of syntax is sometimes crossed in the sense that we also include lexical information if needed. For instance, if we specify the lemma of the concerning preposition in the PP as in *fclass_neg_subj_eff_reflobja_prepobj[um]_verb_neg* (in this case "um" (for); note the encoded reflexive direct object), we leave the pure syntax level.

As mentioned above, one of the goals is the combination of the resources (polarity lexicon and verb annotation). This combination provides us with new target specific sentiment calculations which were not possible in a compositional sentiment analysis purely relying on lexical resources and cannot be reliably inferred via a fuzzy criterion like nearness to other polar words. The effects and expectations of an instantiated syntactic verb pattern in combination with bottom-up propagated and composed polarity can therefore be used to approach the goal of sentence-level sentiment analysis based on a deep linguistic analysis. Furthermore our system offers a possibility to detect violations of expected polarities ("admire a deceitful friend"), i.e., if the bottom-up composed polarity and the effects or expectations coming from the verb frame have an opposite polarity. In our empirical study we wanted to find out whether verb expectations actually are violated in real texts and how reliably those cases could be identified. Our verb resources comprise 305 of these verbs for German, 210 for English and 320 for French.

5 Empirical Investigation

We carried out two different experiments: bi-polar noun phrase level polarity composition and verb-based polarity prediction. Our experiments are based on texts from LeSoir (5'800 articles) and articles from the news platform AgoraVox (4'300 articles), alltogether about 6 million words.

5.1 NP-Level

Our hypothesis was: conflicting polarities in a noun phrase always result in a negative NP, especially we claim that

- a negative adjective reverses or negates the positive polarity of the noun that it modifies.
- a positive adjective functions as intensifier of the negative polarity of the noun that it modifies.

Table 6 gives an overview on the most frequent combinations.

In order to verify this hypothesis, we randomly selected a sample of 20 cases for each of the 6 most frequent combination types from our result. The evaluation consisted of two steps: first, we evaluated whether the noun phrases are overall negative, positive or ambiguous. We attributed the value "yes", "no" or "ambiguous" to each noun phrase. The results are listed in Table 7 below.

Tab. 6: Most frequent combinations.

Adj-Noun Combination	Frequency	Adj-Noun Combination	Frequency
1. A POS , A NEG	1041	4. A POS , J NEG	242
2. A NEG , A POS	691	5. A NEG , J POS	234
3. J POS , A NEG	349	6. J NEG , A POS	155

Tab. 7: Composition results.

Adj-Noun Combination	yes (negative)	no (positive)	ambiguous
A POS , A NEG	14/20 (70%)	4/20 (20%)	2/20 (10%)
A NEG , A POS	17/20 (85%)	2/20 (10%)	1/20 (5%)
J POS , A NEG	12/20 (60%)	4/20 (20%)	4/20 (20%)
A POS , J NEG	18/20 (90%)	0/20 (0%)	2/20 (10%)
A NEG , J POS	17/20 (85%)	2/20 (10%)	1/20 (5%)
J NEG , A POS	19/20 (95%)	0/20 (0%)	1/20 (5%)
Total	97/120 (81%)	12/120 (10%)	11/20 (9%)

According to our manual evaluation, 97 out of 120 selected conflict cases (which corresponds to 81%) validate our hypothesis. Indeed, we can easily identify positive adjectives as intensifiers of their negative head nouns:

- *célèbre catastrophe* "famous catastrophe"
  (`A_POS adjective + A_NEG noun`)
- *glorieuse incertitude* "glorious uncertainty
  " (`J_POS adjective + A_NEG noun`)
- *violation délibérée* "deliberated violation"
  (`A_POS adjective + J_NEG noun`)

The following examples illustrate how negative adjectives act as negators or shifters of their positive head nouns:

- *goût amer* "bitter taste"
  (`A_NEG adjective + A_POS noun`)
- *fausses innocences* "false innocence"
  (`A_NEG adjective + J_POS noun`)
- *ambition cynique* "cynical ambition"
  (`J_NEG adjective + A_POS noun`)

We evaluated 19% of the selected cases as either ambiguous or as contradicting our hypothesis. A contradiction of the hypothesis means that the overall polarity of a noun phrase should have been computed as overall positive instead of overall negative, such as in the following cases:

- *sourire ravageur* "charming smile"
  `(A_NEG adjective + A_POS noun)`
- *bouleversante sincérité* "overwhelming sincerity"
  `(A_NEG adjective + J_POS noun)`
- *lutte antiterroriste* "antiterrorist fight"
  `(J_POS adjective + A_NEG noun)`

Based on our empirical study, we formulate one aggregation rule for each of the 6 analysed combination types:

- If an A POS adjective modifies an A NEG noun, the A POS adjective acts as an intensifier. The overall polarity of the NP is A NEG.
- If an A NEG adjective modifies an A POS noun, the A NEG adjective shifts the positive polarity of the noun. The overall polarity of the NP is A NEG.
- If an J POS adjective modifies an A NEG noun, the J POS adjective adds a further qualification to the noun. The overall polarity of the NP is A NEG.
- If a A POS adjective modifies an J NEG noun, the adjective acts as an intensifier of the noun. The overall polarity of the NP is J NEG.
- If a A NEG adjective modifies an J POS noun, the adjective shifts the noun. The overall polarity of the NP is reversed to J NEG.
- If an J NEG adjective modifies a A POS noun, the meaning and polarity type of the adjective overrule those of the noun. The overall polarity of the NP is J NEG.

Additional knowledge is needed in order to cope with the exceptions discussed above. These rules provide the best solution given our current resources.

5.2 Verb Level

We searched for verb expectation violations, since we believe that they form an interesting phenomenon. They might help identify parts of a text that represent controversial opinions or offending passages. Expectation violations are rare in our French corpus, given 450'000 sentences, only about 500 conflicts were found:

- 410 (81.18% A_NEG or A_POS conflict
- 72 (14.25% J_NEG or J_POS conflict
- 23 (4.55% F_NEG or F_POS conflict

For instance, if the verb has a positive expectation, a A_-F_- or J_NEG might cause a conflict. No clear conclusions can be drawn from our data concerning a general rule for these cases. However, we identified four different verb-specific conflict classes: the fine-grained polarity tags do determine whether a polarity conflict occurs or not, but they behave differently depending on the verbs. The categories and rules that we identified are the following:

1. `A_NEG`, `F_NEG` and `J_NEG` always cause conflicts, except if they are further labelled as `passive`.
   Verbs: alimenter, nourrir
2. `A_NEG` and `F_NEG` do not produce conflicts, except if they are further labelled as `strong`. `J_NEG` words and expressions always generate conflicts.
   Verbs: accepter, accueillir, accorder, adorer, aider, aimer, apprécier, encourager, défendre, permettre, privilégier, prôner, soutenir, suggérer
3. Nor `A_NEG`, `F_NEG` or `J_NEG` can cause conflicts. The verbs have a positive impact on the negative polarity: they diminish the negative polarity.
   Verbs: corriger, soulager
4. `A_NEG`, `F_NEG` and `J_NEG` always generate conflicts because
   (a) the verb intensifies negative expressions.
       Verbs: assurer, conforter, cultiver, favoriser
   (b) the proposition establishes a negative polarity effect or relation with regard to somebody or something.
       Verbs: mériter, désirer, offrir, profiter, promettre

For instance, the fourth category deals with verbs that always generate conflicts, regardless of the type of negative polarity they modify. We divided the verbs into two distinct subcategories (a and b). The verbs listed in the subcategory a), *assurer* "assure", *conforter* "strengthen", *cultiver* "cultivate", *favoriser* "favour", have, as far as our data shows, an intensifying effect when they modify negative expressions. The type of negativity is not relevant:

1. a. *favoriser le désespoir* "favour despair"
   b. *favoriser le terrorisme* "favour terrorism"
2. a. *cultiver le mensonge et la trahison* "cultivate lies and betrayals"
   b. *cultiver la haine* "cultivate hatred"

We have started to further explore verb expectation conflicts for the other languages. Especially for German, where we have a larger corpus (compared to French), namely the DeWaC corpus (see Baroni et al. (2009)) comprising 90 Million sentences. Hopefully, we can get a clearer picture given more data.

6 Related Work

No special attention is paid to bi-polar phrases in the literature. A simple composition rule is used, namely that positive and negative yields negative (e.g. Choi and Cardie (2008)). In this paper, we have had a closer look at these special noun phrases and tried to fix better tailored rules.

The role that verbs play in sentiment analysis is not so broadly acknowledged compared to adjectives and nouns. However, there are a few approaches that strive to clarify the impact of verbs, e.g. Chesley et al. (2006), Neviarouskaya et al. (2009) and Reschke and Anand (2011). Chesley et al. (2006) use verb classes and attach a prior polarity to each verb class. They show how these verb classes contribute as features to the accuracy of their approach. No attention is paid to the verb's arguments. This, however, is the primary focus of the work of Neviarouskaya et al. (2009). In their approach, verb classes are used to specify effects on the grammatical roles of verbs (subject, ...).

An approach that focuses on the interplay between the polarity of the bearers of grammatical roles of a verb and the overall verb frame polarity is Reschke and Anand (2011). Again, verb classes are used, e.g. verbs of having, withholding, disliking and linking. Frame polarity depends on the polarity of the subject and the direct object. E.g. an instantiated verb frame of the verb "lack" is positive, if the subject is negative and the direct objects is positive ("your enemy lacks good luck" is positive). Nothing is derivable about the polarity preferences of these verbs, e.g. that "lack" has positive polarity preference for its direct object. This is what our approach reveals, so these two approaches are complementary.

7 Conclusion

We have introduced fine-grained lexical resources for French, German and English sentiment analysis. In order to properly compose word level polarity to phrase level polarity and finally clause level polarity the role of bipolar phrases needs to be clarified. We have carried out experiments with a French corpus in order to develop such composition rules.

We also have introduced novel verb resources, where verbs have effects and expectations on their arguments. Such a resource is useful in order to fix the contextual polarity of neutral noun phrases occuring as an argument of these verbs. They, so to speak, inherit the polarity expectation or effects. In this paper, we have, however, focussed on conflicts arising from a top down restriction that gets violated bottom up. Further work is needed in order to clarify how to deal with such violations. Our study suggests that it is verb-specific, but that it depends on our fine-grained polarity categories as well. This is future work. Detecting such violations might enable our sentiment analysis system to detect interesting text passages (e.g. controversial stance).

Bibliography

Baroni, M., Bernardini, S., Ferraresi, A., and Zanchetta, E. (2009). The WaCky Wide Web: A collection of very large linguistically processed web-crawled corpora. In *Language Resources and Evaluation (LREC)*, pages 209–226.

Chesley, P., Vincent, B., Xu, L., and Srihari, R. K. (2006). Using verbs and adjectives to automatically classify blog sentiment. In *Proc. AAAI-2006 Spring Symposium on Computational Approaches to Analyzing Weblogs*, pages 27–29.

Choi, Y. and Cardie, C. (2008). Learning with compositional semantics as structural inference for subsentential sentiment analysis. In *Proc. of EMNLP*.

Maks, I. and Vossen, P. (2012). A lexicon model for deep sentiment analysis and opinion mining applications. *Decision Support Systems*, 53(4):680–688.

Martin, J. R. and White, P. R. R. (2005). *Appraisal in English*. Palgrave, London.

Moilanen, K. and Pulman, S. (2007). Sentiment composition. In *Proc. of RANLP-2007*, pages 378–382, Borovets, Bulgaria.

Neviarouskaya, A., Prendinger, H., and Ishizuka, M. (2009). Semantically distinct verb classes involved in sentiment analysis. In Weghorn, H. and Isaías, P. T., editors, *IADIS AC (1)*, pages 27–35. IADIS Press.

Reschke, K. and Anand, P. (2011). Extracting contextual evaluativity. In *Proc. of the Ninth Intern. Conf. on Computational Semantics*, pages 370–374.

VISL-group (2013). *VISL CG-3. http://beta.visl.sdu.dk/cg3.html*. Institute of Language and Communication (ISK), University of Southern Denmark.

Michaela Regneri and Diane King

Automatically Evaluating Atypical Language in Narratives by Children with Autistic Spectrum Disorder

Abstract: We conducted an automatic analysis of transcribed narratives by children with autistic spectrum disorder. Compared to previous work, we use a much larger dataset with control groups that allow for the distinction between delayed language development and autism-specific difficulties, and that allows differentiation between episodic and generalized narratives. We show differences related to expression of sentiment, topic coherence and general language level. Some features thought to be autism-specific present as delayed language development.

1 Introduction

In recent years, there has been a growing interest in automated diagnostic analysis of texts which detects idiosyncratic language caused by certain mental conditions: while standard approaches to text classification mostly focus on pure identification of certain users or user groups, often with hardly interpretable features (like function words), diagnostic analysis has the additional goal of making sense out of the actual features. The hope here is to gain more insight into the underlying disorder by analysing how it affects language.

Our focus is on the diagnostic analysis of texts by adolescents with autistic spectrum disorder (ASD). ASD is a neurodevelopmental disorder characterised by impairment in social interaction and communication. Whilst there has been much research on ASD, there are not many corpus analyses to support assumptions on language development and ASD. In general, it is difficult to gather large datasets with meaningful control groups. Previous approaches employed small datasets of very restricted domains like retellings [14, 18].

Our work is based on the largest available dataset with narratives told by teenagers with ASD [11]. In contrast to other datasets, this has carefully matched

Michaela Regneri: Dept. of Computational Linguistics, Saarland University, Saarbrücken, Germany, e-mail: michaela_regneri@spiegel.de
Diane King: National Foundation for Educational Research, London, United Kingdom, e-mail: d.king@nfer.ac.uk

control groups, which enables the distinction of delayed language development from ASD-specific difficulties. We analyse this data according to different features, which are in line with theories on ASD and language, and which have partially been used in previous work on texts from children with ASD (like term frequency × inverse document frequency). We (re-)evaluate all these features and compare their prevalence in the ASD group with two control groups, and show where we can find meaningful differences – and where we cannot. The result is a survey on differences of the language of adolescents with ASD and their typically developing peers.

We first review previous research on ASD-specific language and related work on automated analysis (Sec. 2). We then describe the dataset and, based on this, our research objectives (Sec. 3) before presenting the results of our data analysis (Sec. 4). After a resumee on the results (Sec. 5) we conclude with a summary and suggestions for future work (Sec. 6).

2 Related Work

Many automated approaches to diagnostic analysis detect Alzheimer's and related forms of dementia [8, 12]. Some classifiers are capable of automated diagnosis from continuous speech [1], or even predicting the cognitive decline decades before its actual onset [15].

Other systems recognize spontaneous speech by individuals with more general mild cognitive impairments, for adults [16] and also for children [7]. [9] present an unusual study on the language of adult patients with schizophrenia.

Previous research on narratives of children with ASD has reported difficulties with both structural and evaluative language. Individuals with ASD have difficulties in expressing sentiment, and make fewer references to mental states [4, 20]. However, other research shows, when carefully matched with comparison groups, many of these differences are not evident.

More basic problems emanate from a general lower syntactic complexity [21]. Subjects also seem to struggle with discourse coherence [13], which may, in part, be related to difficulties with episodic memory [2].

Some of these language difficulties have been subject to automated analysis: [14] analyzed data of very young children (6–7 years old). They built an automated classifier that distinguishes sentences uttered by children with ASD from sentences of two control groups (one with children with a language-impairment, one with typically developing children). The authors themselves note some drawbacks of their underlying dataset, in particular that some children in the ASD group were also classified as language-impaired. In consequence, a clear distinction between these groups was impossible.

In a follow-up study, [18] analysed whole narratives (retellings) told by children (mean age 6.4) with ASD compared to a typically developing control group (with the same average age and IQ). They used the $tf \times idf$ measure to identify idiosyncratic words uttered by children in the ASD group. The texts from the control group and some crowdsourced retellings from typically developing adults served as a basis for determining unusualness.

We consider this work of Rouhizadeh et al. as a basis and want to move further with a different dataset and different features: First, the dataset we are using [11] is much larger than any of the other automatically analysed ASD-datasets. Furthermore, the age range of the participants (11–14 years) is more suitable for this study, because narrative abilities are not fully in place before the age of nine years [10]. Additionally, this dataset allows us to properly distinguish ASD-specific features from language development issues. Finally, we want to add features concerning references to mental states.

As our main objective, we do not aim to build a classifier that finds stories by children with ASD, but rather want to examine which automatically derived features can further our understanding of language and ASD, and which (presumably relevant) features might question existing theories.

3 Corpus & Analysis Objectives

In the following, we will describe the dataset and why it is suited for this task, and then summarize the core techniques and objectives of our analysis.

3.1 A Dataset of Narratives by Children with ASD

We base our analysis on an existing dataset [11]. The corpus contains transcripts of narratives about 6 everyday scenarios: SPENDING FREE TIME, BEING ANGRY, GOING ON HOLIDAYS, HAVING A BIRTHDAY, HALLOWEEN and BEING SCARED.

The subjects were divided in three groups: 27 children with ASD, one comparison group of 27 children matched by chronological age and nonverbal ability, and another of 27 children individually matched on a measure of expressive language and on nonverbal ability. All groups had average scores on nonverbal and verbal measures, as measured by the Matrices test of the BAS II and the BPVS II. The average age difference between the language-matched control group and the two other groups is 17 months.

Each child told two stories for each scenario: first a *specific narrative* about a particular instance (answering a prompt like “Can you tell me about a time when you went on holiday?”), and second a *general narrative* that contains a script-like prototypical scenario description (“What usually happens when someone goes on holiday?”). This distinction of general and specific narratives allows us to assess both the episodic memory of the children and their ability to generalize to commonsense knowledge. Each of the 27 subjects told 2 × 6 stories which sums up to 972 narratives (324 per group).

King et al.transcribed all texts and added annotations coding references to mental states, causal statements, negative comments, hedges, emphatic remarks and direct speech. Table 1 shows some basic statistics on the corpus. The figures distinguish the subject groups and also the general (GEN) and specific (SEN) event narratives. The narratives from the ASD group were, in general, shorter, both with respect to the number of utterances and the number of words per utterance. We also find fewer references to mental states for the general narratives, which will later be an anchor point for our own analyses.

Tab. 1: Text and utter. length & mental state references, by group and story type.

	utterances / text		words / utterance		mental states	
	GEN	SEN	GEN	SEN	GEN	SEN
ASD	28.9	33.4	8.5	8.5	0.2	0.3
Language Match	24.3	38.4	13.5	14.1	0.4	0.3
Age Match	24.7	39.2	15.4	13.3	0.5	0.4

This dataset is particularly well suited to analyse difficulties with coherence (because it contains whole narratives), expression of sentiment (because it contains emotion-centered scenarios), and the distinction of language development issues from ASD-specific difficulties (because it has a language-matched control group). We further want to make use of the differences for general and specific narratives, because this distinction offers new perspectives on more general cognitive capabilities behind the language in children with ASD.

3.2 Automatic Analysis of ASD-Specific Language

Our goal is to look for evidence in the data that heuristically tests theories on ASD and language. The following details the theoretical issues underpinning this and the corresponding automatically retrievable features we want to explore:

1. **General language difficulties:** As a first step, we show features potentially indicating general language development difficulties in the ASD group: we assess the use of low-frequency words and idiosyncrasies in pronoun use.
2. **Difficulties with topic coherence:** We want to re-visit the use of unusual words and the degree of their unusualness by subjects with ASD, using *tf* × *idf* (like Rouhizadeh et al.). We consider words as unusual if they are off-topic, i.e. deviate considerably from the current narrative focus.
3. **Difficulties with verbalizing sentiments:** Individuals with ASD are known to have difficulties with references to mental states. We use a state-of-the-art sentiment analyser to compare which kind of sentiments are expressed most frequently by which subject group. Further, we generally consider the scenarios of being SCARED and ANGRY as potential source of difficulties.
4. **Differences between recalling episodes and generalizing scenarios:** Our dataset contains scenario-matched general (script-like) and specific (episodic) narratives, which allows us to evaluate which of the two are more difficult and in which respect. This is particularly interesting because tasks related to episodic memory as well as the development of generic common sense knowledge are, in theory, challenging for individuals with ASD.
5. **Differences between language development issues and ASD-specific symptoms:** Language-related problems of children with ASD are hard to classify as either general language-development problems or as ASD-specific. Because our dataset contains a control group which is matched for language capabilities, we can examine this distinction.

4 Analysis of Linguistic Features

We evaluate different linguistic features on the dataset, which can indicate different language idiosyncrasies for ASD: We apply some measures for the general state of language development (Sec. 4.1), difficulties with topic coherence (Sec. 4.2), and the expression of sentiments (Sec. 4.3).

4.1 General Language Competence

The language competence of the children with ASD is reported to be in the average range and to match the respective control group. We try to validate those results, first by assessing the use of low-frequency words, which is associated with high language competence. Second, we analyse pronoun use: both a generally low pronoun use and an over-proportional use of first person singular pronouns indicate earlier stages of language development [6].

Low-Frequency Words

To measure word frequencies, we mainly employ Kilgariff's frequency list[1] drawn from parts of the BNC corpus [5]. We have chosen this particular corpus because it includes both written language and transcribed speech, and because it is reasonably large (100M source tokens).

Evaluating the average word (type) frequency per narrative does not yield any difference between the subject groups. (We also tested word and bigram frequencies from google n-grams [3], with the same result.) We also show a more focused analysis on low-frequency words: we count low-frequency words according to different frequency thresholds. E.g. if we set a threshold of 0.01%, we measure the proportion of tokens from a type which contributes less than 0.01% to the tokens in the reference corpus.

Table 2 shows the results, listed by subject group and frequency threshold. While the proportion of low-frequency words is basically equal in the age-matched control group and the (younger) language-matched control group, it is slightly lower in the narratives from the ASD group. This is only partially explainable by a lower language-competence of the ASD group, in particular given the equal results for both control groups.

Tab. 2: Token proportions below small frequency thresholds (w.r.t. the source corpus).

Group	< **0.00003**%	< **0.0001**%	< **0.001**%	< **0.01**%
ASD	**0.18**	**0.26**	**0.45**	**0.57**
Language Match	0.23	0.31	0.52	0.65
Age Match	0.22	0.31	0.52	0.66

Pronoun Use

As indicators of language development level, we analyse the overall proportion of pronouns, and the proportion of first person singular pronouns (1psg, *I*, *my* etc.). In general, pronoun use *increases* with language development, while the use of 1psgs *decreases* during language development.

1 http://www.kilgarriff.co.uk/bnc-readme.html

Table 3 shows the results: column 2 indicates the overall (token-based) proportion of pronouns (relative to all word tokens). The following columns show the proportion of 1psgs, relative to all pronouns; col. 3 shows the overall average, col. 4 & 5 split the numbers up by narrative type (specific vs. generic).

Tab. 3: Overall pronoun use & Proportion of first person singular pronouns, also contrasting general event narratives (SEN) and specific event narratives (GEN).

Subject group	pronouns / words	1psg / pronouns	1psg SEN	1psg GEN
ASD	0.08	0.49	0.75	**0.23**
Language Match	0.10	0.47	0.76	0.18
Age Match	0.10	0.46	0.75	0.17

The results show that there are differences in pronoun use, but that these appear to have other reasons than differences in language ability: on average, all three groups show comparable results. However, there is an over-proportional use of 1psgs in the general narratives from the ASD group. This might indicate that the difficulties of the children with ASD are not due to language capabilities, but rather to other systematic problems.

As shown in table 4, those differences in general narratives are not equal for all scenarios. In many cases, the ASD group shows no differences to their language-matched controls (SCARED) or even their age-matched controls (BIRTHDAY, ANGRY). However, event-driven topics (FREE TIME, HALLOWEEN) seem to cause more difficulties in generalizing away from the first person perspective.

Tab. 4: Proportion of 1st person singular pronouns in general narratives, by scenario.

Group	FREE TIME	BEING SCARED	BIRTH-DAY	HOLIDAY	HALLO-WEEN	BEING ANGRY
ASD	**0.51**	0.22	0.20	0.18	**0.18**	0.09
Language Match	0.33	0.23	0.32	0.11	0.02	0.09
Age Match	0.36	0.13	0.21	0.13	0.09	0.10

4.2 Topic Coherence

Another commonly described phenomenon in children with ASD is difficulty with topic maintenance. [18] describe how to use $tf \times idf$ to detect idiosyncratic words in texts by children with ASD. We wanted to validate their underlying assumption that children with ASD in fact use more out-of-topic words than typically developing children.

Unusual Off-Topic Words

Term Frequency / Inverse Document Frequency [17, $tf \times idf$] is an established measure in information retrieval to quantify the association of a term t with a document d_j from a document collection $D = \{d_1, d_2, ...d_n\}$. We compute the tf and idf, as follows, with $freq(t, d)$ as the number of $t's$ occurrence in d:

$$tf(t, d) = 0.5 + \frac{0.5 \times freq(t, d)}{max\{freq(w, d) : w \in d\}} \qquad idf(t, D) = log \frac{|D|}{|\{d \in D : t \in d\}|}$$

Intuitively, a term t has a high $tf \times idf$ score for a document d if t is very frequent in d, and very infrequent in all other documents. The division by the maximum frequency of any term in d normalizes the result for document length. We compute $tf \times idf$ for each narrative with all narratives of the same scenario as underlying document collection. This is similar to the experiment by [18], with the additional normalization for text length, and with less narrow topics.

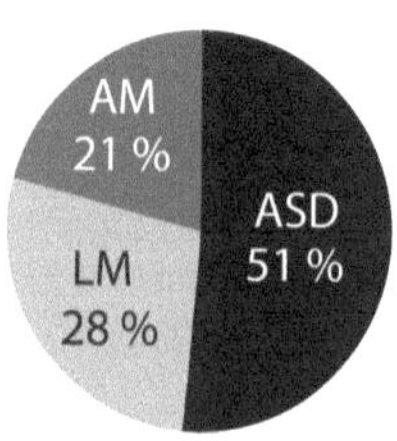

Fig. 1: Distribution of the 1% most unusual words according to $tf \times idf$ scores.

Table 5 shows the results. There were no indicative differences on general and specific narratives, so we omit the distinction ($tf \times idf$ was always higher for specific narratives). On average, there are only small absolute differences between the ASD group and both control groups. The maximal scores are the same for both the ASD

Tab. 5: *tf* × *idf* scores detailed by scenario; maximum scores are on the left, average scores on the right. *LM* and *AM* mark language and age match, respectively.

Scenario	Maximum *tf* × *idf*			Average *tf* × *idf*		
	ASD	LM	AM	ASD	LM	AM
SPENDING FREE TIME	5.15	5.15	3.86	1.44	1.38	1.37
BEING SCARED	5.16	5.16	4.73	1.66	1.64	1.63
HAVING A BIRTHDAY	4.47	3.87	4.30	1.49	1.39	1.40
GOING ON HOLIDAY	5.14	3.85	3.60	1.42	1.37	1.40
HALLOWEEN	5.16	5.16	4.47	1.57	1.42	1.44
BEING ANGRY	5.15	5.15	4.50	1.70	1.60	1.55
AVERAGE	5.16	5.16	4.73	1.53	1.46	1.46

group and their language-matched controls, which indicates the strong relationship of *tf* × *idf* with lexical knowledge development.

Comparing scenarios, we find some wider topics with fewer "usual" words (SCARED and ANGRY). For the more episodic narratives like BIRTHDAY and HALLOWEEN, the difference between the ASD group and the control groups is clearer.

Whilest the overall picture does not show the expected difference, words with the highest *tf* × *idf* occur more frequently in the ASD group: Figure 1 shows the distribution of the 1% words with the highest *tf* × *idf* scores. More than half of those outstandingly unusual words come from the ASD group, while the other half is divided in comparable pieces between the two control groups.

4.3 Sentiment & References to Mental States

Our dataset confirms that conveying emotions presents more difficulties for children with ASD than for their typically developing peers. Of particular interest are the narratives that specifically point to the emotionally defined situations of BEING SCARED and ANGRY. The mere frequency of references to mental states is reported to be is much lower in the ASD group than in any of the other two groups [11] We want to test whether we can confirm such an analysis with scalable automated means, and additionally evaluate the type and distribution of emotions conveyed. We use one of the best automatic sentiment analysers [19], which evaluates sentences as either neutral, positive or negative.

Table 6 shows that there are considerably more neutral sentences in the narratives of the ASD group than in the control groups. Further, they contain very few negative sentences but a similar proportion of positive sentences. Table 7 shows that references to the mental states of other people seems to be most difficult for

Tab. 6: Proportions of negative, neutral and positive sentences in the dataset.

Sentiment	ASD	LM	AM
negative	*0.37*	0.54	**0.60**
neutral	**0.52**	0.36	*0.29*
positive	0.11	0.10	0.11

Tab. 7: Proportion of neutral sentences in general narratives.

Scenario	ASD		Lng. match		Age Match	
	GEN	SEN	GEN	SEN	GEN	SEN
FREE TIME	**0.63**	**0.48**	0.41	0.25	0.38	0.28
BE SCARED	**0.65**	**0.51**	0.41	0.32	0.30	0.26
BIRTHDAY	0.54	0.44	0.43	0.31	0.24	0.30
HOLIDAY	0.56	0.33	0.41	0.34	0.33	0.35
HALLOWEEN	0.51	0.45	0.44	0.34	0.29	0.26
BE ANGRY	**0.64**	**0.48**	0.39	0.26	0.32	0.20

the children with ASD, with the emotion-defined scenarios (SCARED & ANGRY), showing the highest proportion of neutral statements.

5 Discussion

Our evaluation is an initial exploration, and more research needs to be done to arrive at a final analysis suite for ASD-related language. The results we found do not provide clear answers with respect to all of our original research objectives. Nevertheless, we found some interesting first evidence:

The use of **low-frequency words** is slightly less common in the ASD group, while the results for the two control groups are very similar. This is interesting because the feature, at first glance, looks like an indicator of language deficiency, but on further examination seems to be a by-product of other difficulties specific to the ASD group (e.g. a general difficulty of describing certain scenarios).

Contrary to our expectation, we did not find any differences in the overall **pronoun use** (indicating text coherence). With first person pronouns, generalization over episodic scenarios is harder for children with ASD.

The picture for **topic coherence** is less clear than expected. [18] also used $tf \times idf$, but did not provide evidence that this is a distinctive feature. Contrary to their basic assumption, words with high $tf \times idf$ are not something exclusive to children with ASD, but rather occur in all groups. Comparing maximal $tf \times idf$ scores also raises the question as to whether the results are actually underpinned by language deficiencies. What *can* be shown is that the words with the highest $tf \times idf$ are twice as likely to stem from a child with ASD.

Our clearest evidence concerns **sentiment and references to mental states.** Whilst it was clear from the corpus annotation that children with ASD use fewer direct references to mental states, we did not know before this analysis about differences of emotion types and differences between scenarios. Automated sentiment analysis gave first answers to both: while we have no differences between the subject groups in the use of positive sentences, there are far more neutral sentences in the ASD group, and far fewer negative ones. Further, the differences are scenario-dependent, and are particularly striking for the "emotional" scenarios, which require references to mental states by default.

6 Conclusion

We have presented an explorative analysis of narrative language produced by children with ASD. In contrast to previous approaches, we used a much larger dataset that allows for the distinction between language deficiencies and ASD-specific symptoms. We further applied a wider diversity of automated analyses, and showed that some features (like unusual words as classified by $tf \times idf$) are not necessarily very distinctive for narratives produced by children with ASD.

What we did not provide is an actual classifier for the automatic distinction between texts by children with ASD and texts produced by one of the control groups. One reason for this is that although some very shallow features (like text length) would yield a very good result on our datasets, rather than finding the most distinctive features, we wanted to find the most telling features which show where the actual language difficulties for children with ASD are. Classifying according to length e.g. would only summarize the outcome of all those difficulties in one rather meaningless classification metric.

We also did not discriminate between individuals who may be on different points of the spectrum, because the dataset does not account for this. The diagnosis of autism was not made by the researchers but by clinicians and we have no information other than that the child was 'high functioning' or had Asperger Syndrome. Most had a diagnosis of high functioning autistic spectrum disorder. A more fine-grained analysis could be an an interesting area for future research.

In future work, we want to explore more features and details of the dataset. One direction would be to more closely examine topic models and the nature of content they actually produce in order to gain more information on topic coherence. We also want to explore different text types (such as fictional narrative), and also use texts by adults to evaluate if, and to what degree, language difficulties persist in adulthood.

Acknowledgements

We want to thank Julie Dockrell and Morag Stuart for their advice on collection of the original data. We also thank the anonymous reviewers for their helpful comments. – The first author was funded by the Cluster of Excellence "Multimodal Computing and Interaction" in the German Excellence Initiative.

Bibliography

[1] Baldas, V., Lampiris, C., Capsalis, C., Koutsouris, D.: Early diagnosis of alzheimer's type dementia using continuous speech recognition. Lecture Notes of the Institute for Computer Sciences, Social Informatics and Telecommunications Engineering, vol. 55, pp. 105–110. Springer Berlin Heidelberg (2011)

[2] Boucher, J., Bowler, D.: Memory In Autism: Theory and Evidence. Memory in Autism: Theory and Evidence, Cambridge University Press (2010), http://books.google.de/books?id=ExOccQAACAAJ

[3] Brants, T., Franz, A.: Web 1t 5-gram version 1. Linguistic Data Consortium (2006)

[4] Capps, L., Losh, M., Thurber, C.: "the frog ate the bug and made his mouth sad": Narrative competence in children with autism. Journal of Abnormal Child Psychology 28(2), 193–204 (2000), http://dx.doi.org/10.1023/A%3A1005126915631

[5] Clear, J.H.: The digital word. chap. The British National Corpus, pp. 163–187. MIT Press, Cambridge, MA, USA (1993), http://dl.acm.org/citation.cfm?id=166403.166418

[6] Forrester, M.: The Development of Young Children's Social-cognitive Skills. Essays in developmental psychology, Lawrence Erlbaum Associates (1992), http://books.google.de/books?id=ABp6Tv-fl-sC

[7] Gabani, K., Sherman, M., Solorio, T., Liu, Y., Bedore, L., Peña, E.: A corpus-based approach for the prediction of language impairment in monolingual english and spanish-english bilingual children. In: Proceedings of NAACL-HLT 2009 (2009)

[8] Hirst, G., Wei Feng, V.: Changes in style in authors with alzheimer's disease. English Studies 93(3) (2012)

[9] Hong, K., Kohler, C.G., March, M.E., Parker, A.A., Nenkova, A.: Lexical differences in autobiographical narratives from schizophrenic patients and healthy controls. In: Proceedings of EMNLP-CoNLL 2012 (2012)

[10] Karmiloff-Smith, A.: Language and cognitive processes from a developmental perspective. Language and Cognitive Processes 1(1), 61–85 (1985), http://www.tandfonline.com/doi/abs/10.1080/01690968508402071

[11] King, D., Dockrell, J.E., Stuart, M.: Event narratives in 11–14 year olds with autistic spectrum disorder. International Journal of Language & Communication Disorders 48(5) (2013)
[12] Le, X., Lancashire, I., Hirst, G., Jokel, R.: Longitudinal detection of dementia through lexical and syntactic changes in writing: a case study of three british novelists. Literary and Linguistic Computing 26(4) (2011)
[13] Loveland, K., Tunali, B.: Narrative language in autism and the theory of mind hypothesis: a wider perspective. In: Baron-Cohen, S., Tager-Flusberg, H., Cohen, D.J. (eds.), Narrative language in autism and the theory of mind hypothesis: a wider perspective. Oxford University Press (1993)
[14] Prud'hommeaux, E.T., Roark, B., Black, L.M., van Santen, J.: Classification of atypical language in autism. ACL HLT 2011 p. 88 (2011)
[15] Riley, K.P., Snowdon, D.A., Desrosiers, M.F., Markesbery, W.R.: Early life linguistic ability, late life cognitive function, and neuropathology: findings from the nun study. Neurobiology of Aging 26(3) (2005)
[16] Roark, B., Mitchell, M., Hosom, J., Hollingshead, K., Kaye, J.: Spoken language derived measures for detecting mild cognitive impairment. Audio, Speech, and Language Processing, IEEE Transactions on 19(7), 2081–2090 (2011)
[17] Robertson, S.E., Sparck Jones, K.: Document retrieval systems. Taylor Graham Publishing (1988)
[18] Rouhizadeh, M., Prud'hommeaux, E., Roark, B., van Santen, J.: Distributional semantic models for the evaluation of disordered language. In: Proceedings of NAACL-HLT 2013 (2013)
[19] Socher, R., Perelygin, A., Wu, J., Chuang, J., Manning, C.D., Ng, A.Y., Potts, C.: Recursive deep models for semantic compositionality over a sentiment treebank. In: Proceedings of EMNLP 2013 (2013)
[20] Tager-Flusberg, H.: Brief report: Current theory and research on language and communication in autism. Journal of Autism and Developmental Disorders 26(2) (1996)
[21] Tager-Flusberg, H., Sullivan, K.: Attributing mental states to story characters: A comparison of narratives produced by autistic and mentally retarded individuals. Applied Psycholinguistics 16 (1995)

Michael Muck and David Suendermann-Oeft

How to Make Right Decisions Based on Currupt Information and Poor Counselors

Tuning an Open-Source Question Answering System

Abstract: This paper describes our efforts to tune the open-source question answering system OpenEphyra and to adapt it to interface with multiple available search engines (Bing, Google, and Ixquick). We also evaluate the effect of using outdated test data to measure performance of question answering systems. As a practical use case, we implemented OpenEphyra as an integral component of the open-source spoken dialog system Halef.

1 Introduction

Over the last years, academic and industrial interest in automatic question answering technology has substantially grown [1]. In addition to commercial implementations question answering engines such as IBM's Watson DeepQA [2] or Wolfram Alpha [3], there is a number of academic engines such as QA-SYS [4], OSQA [5], and OpenEphyra [6]. Being an open-source software, the latter turned out to be particularly suitable for a number of applications the DHBW Spoken Dialog Systems Research Center is working on, such as the free spoken dialog system Halef [7].

To make sure OpenEphyra meets the quality standards required for an integration into Halef, we undertook a thorough quantitative assessment of its performance. In order to do so, we used a standard test set for question answering (the NIST TREC-11 corpus) which raised a number of issues with respect to the consistency of test sets in the question answering domain, discussed further in Section 2.

Even though OpenEphyra is a free software, its original implementation was based on Bing API [8], a commercial service provided by Microsoft. To become independent of a certain provider, we implemented APIs to interface with a number

Michael Muck: DHBW Stuttgart, Stuttgart, Germany and Tesat-Spacecom GmbH & Co. KG, Backnang, Germany
David Suendermann-Oeft: DHBW Stuttgart, Stuttgart, Germany

of regular web search engines (Google, Bing, Ixquick). One of our major interests was to understand how performance depends on the specific search engine used and whether system combination would result in performance gain. We also analyzed the impact of multiple parameters, for instance associated with the number of queries per question or the number of search results taken into account to compile the final system response. These activities are described in Section 3.

Finally, we describe our efforts to embed OpenEphyra as a webservice used as component of the dialog manager in the spoken dialog system Halef, a telephony-based distributed industry-standard-compliant spoken dialog system in Section 4.

2 Reflections on the Test Set

2.1 Measuring Performance of Questions Answering Systems

A common method to measure the performance of a question answering system is to use a test corpus of a certain number of questions each of which is associated with a set of possible correct (canonical) answers[1]. The system will produce a first-best response for all the involved questions which are then compared to the set of canonical answers. If, for a given question, the response is among the canonical answers, this event is considered a match. Ultimately, the total number of matches is divided by the total number of questions resulting in the questions answering accuracy [9].

For the current study, we used the NIST TREC-11 corpus (1394–1893) [10], a standard corpus provided by the National Institute of Standardization and Technology which had been used by IBM's statistical question answering system [11].

Corpus statistics are given in Table 1.

Tab. 1: Statistics of the NIST TREC-11 corpus (1394–1893).

#questions	500
avg #answers	1.06
#questions w/o answer	56

1 There can be multiple ways to define this set including a simple list, regular expressions, or context-free grammars

2.2 Missing the Correct Answer

It is perceivable that, at times, the set of canonical answers does not contain the exact wording of system response which turns out to be correct. In such a case, it will be erroneously counted as an error negatively affecting accuracy. Even worse, in contrast to most other classification problems where the ground thruth is valid once and forever, the answer to certain questions is time-dependent (for example when asking for the location of the next World Cup). A detailed analysis of possible reasons for this lack is given in the following.

Time dependence. Answers may be obsolete:

Example 1. *"Who is the governor of Colorado?"*
- John Hickenlooper
- Bill Ritter

Missing answers. There might be a multitude of terms referring to the same phenomenon:

Example 2. *"What is the fear of lightning called?"*
- astraphobia
- astrapophobia
- brontophobia
- keraunophobia
- tonitrophobia

Scientific ambiguity. Different studies may provide different results:

Example 3. *"How fast does a cheetah run?"*
- 70 mph (*discovery.com*)
- 75 mph (*wikipedia.com*)

Degree of detail. Some questions do not clearly specify how detailed the answer should be. In human interaction, this issue is resolved by means of a disambiguation dialog, or avoided by taking the interlocutor's context awareness into account:

Example 4. *"How did Eva Peron die?"*
- death
- desease
- cervical cancer

Example 5. *"Where are the British Crown jewels kept?"*
- Great Britain
- London
- Tower of London

Different units. There may be differences in physical units (e.g. metrical vs. US customary systems)

Example 6. *"How high is Mount Kinabalu?"*
- 4095 meter
- 4.095 kilometer
- 13,435 feet

Numerical precision. Not only when asking for irrational numbers such as pi, the precision of numerical values needs to be accounted for:

Example 7. *"What is the degree of tilt of Earth?"*
- 23.439 degrees
- 23.4 degrees
- 24 degrees

Partial answers. Some answers consist out of more than just one word, but for recognizing if this answer is correct there is no need for specifying all parts. E.g., when it comes to person names, it is often acceptable to return only certain parts:

Example 8. *"Who was the first woman to run for president?"*
- Victoria Claflin Woodhull
- Victoria Woodhull
- Victoria
- Woodhull

2.3 Effect on the TREC-11 Corpus

After taking a close look at the TREC-11 test set, we had to rectify the set of canonical answers of as many as 120 of the 500 questions, that is about one quarter. To provide an example figure: The described corpus rectification improved one OpenEphyra test configuration (Bing1q) from 37.6% (188/500) to 55.8% (269/481) accuracy.

Obviously, the results of the present study cannot be directly compared to publications of prior publications using TREC-11. However, this would not have been possible anyway given the aforementioned phenomenon of obsolete answers. As a consequence, generally, evaluation of accuracy of question answering systems is a time-dependent undertaking, in a way comparable to measuring performance of, say, soccer teams.

3 Evaluation

After providing a short overview about the architecture of OpenEphyra, details of our enhancements and tuning results are presented.

3.1 A Brief Overview of OpenEphyra's Architecture

Given a candidate question, OpenEphyra parses the question structure and transforms it into what resembles skeletons of possible statements containing the sought-for answers. These skeletons are referred to as *queries*. For instance:

question:	When was Albert Einstein born?
answer:	Albert Einstein was born in X
	Albert Einstein was born on X.

At the same time, the answer type is identified (in this above case, it is a *date*). Then, OpenEphyra searches for documents matching the queries using a search API (OpenEphyra 2008-03-23 used the Bing API). After extracting matching strings from the documents and isolating the respective response candidates (X), a ranking list is established based on the count of identical candidates, metrics provided by the search API, and other factors which, combined in a certain fashion, constitute a confidence score. Finally, the first best answer is returned along with its confidence score [12]. Figure 1 provides a schema of the described process. The initial implementation of OpenEphyra was based on the Bing API requiring a subscription with a limit of 5000 gratis queries per month. This was the main motivation why we implemented interfaces for communication with regular search engines (Google, Bing, and Ixquick). Drawback of these APIs is that, in contract to the official Bing API, they do not have access to the search engines' confidence scores but only to the order of search results. Whereas Google and Bing, limit the number of requests per time unit to a certain degree to prevent abuse by webcrawlers, Ixquick turned out to be rather tolerant in this respect which is why it became our tool of choice.

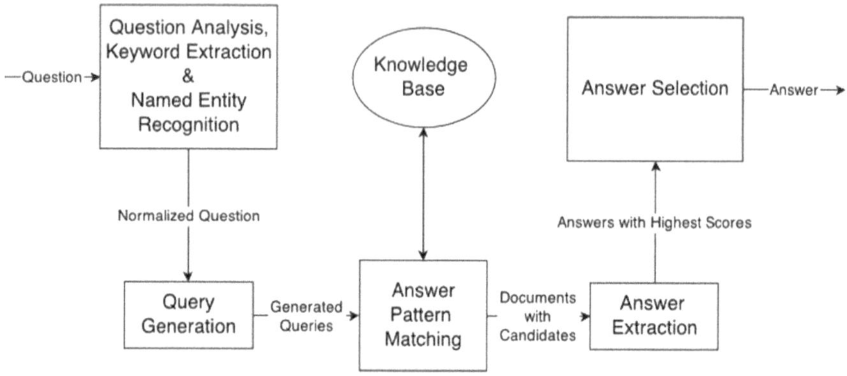

Fig. 1: OpenEphyra's principle architecture.

3.2 Search Engines, Number of Queries, and Number of Documents

First and foremost, we sought to find out which impact the use of web search engines has when compared to the native Bing API. When testing the latter against the TREC-11 corpus, we achieved a benchmark of 57.2% accuracy. It should be noted that OpenEphyra's default settings do not limit the number of queries per question. That is, depending on the question type, a large number of queries might be generated, all of which will be executed. On average, OpenEphyra produces 7.7 queries per question on the TREC-11 corpus. Furthermore, by default, 50 documents are retrieved per query. Table 2 shows accuracy results of multiple combinations of search engines, maximum number of queries, and number of retrieved documents. As aforementioned, due to the limited number of total queries provided by the Google and Bing engines, we were unable to increase the number of queries per question to more than two and the number of documents to more than ten.

Looking at the native Bing API as well as at Ixquick, it can be observed that increasing the number of queries per question from 1, 2, or 3 to unlimited (∞) has a positive impact on accuracy which, however, is not found to be statistically significant on the TREC-11 set (minimum p-value 0.25). Clearly more significant is the impact of the number of documents per question. As an example, Figure 2 shows this relationship for the Ixquick engine. The maximum number of queries per question was unlimited. The accuracy improvement from 10 to 200 considered documents per query was significant with a p-value of 0.08.

Tab. 2: Experimental results.

ID	engine	#queries	#documents	#correct	accuracy/%
BingAllq	Bing	∞	50	275	57.2
Ixquick200	Ixquick	∞	200	270	56.1
Bing1q	Bing	1	50	269	55.9
Ixquick100	Ixquick	∞	100	267	55.5
Bing3q	Bing	3	50	265	55.1
Bing2q	Bing	2	50	263	54.7
Ixquick50	Ixquick	∞	50	258	53.6
Ixquick20	Ixquick	∞	20	253	52.6
Google2q	Google	2	10	247	51.4
IxquickAllq	Ixquick	∞	10	243	50.5
Ixquick1q	Ixquick	1	10	235	48.9
Google1q	Google	1	10	233	48.4
Ixquick2q	Ixquick	2	10	225	46.8
BingW1q	BingW	1	10	202	42.0
BingW2q	BingW	2	10	202	42.0

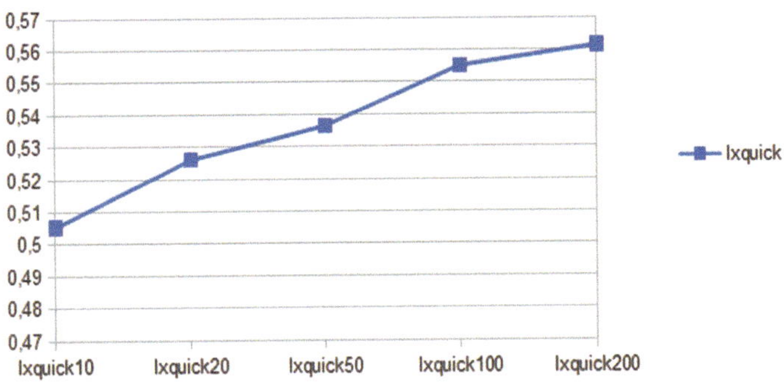

Fig. 2: Dependency of accuracy on the number of retrieved documents.

3.3 Answer Type Dependence

As described in Section 3.1, OpenEphyra generates queries depending on the detected answer type. Hence, we were interested to see how question answering performance depends on the answer type. Figure 3 shows the average performance of Open Ephyra over all test settings mentioned per each of the five main answer types.

- location,
- number,
- names (person/company/nickname/group),
- date,
- rest.

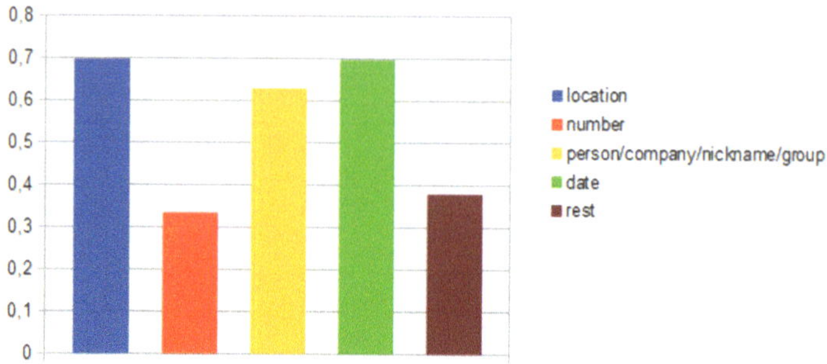

Fig. 3: Dependency of performance on answer types.

While location, names, and date perform pretty decently (60 to 70%), number and other questions perform at less than 40% accuracy. The reason for this are, among others, issues discussed in Section 2.2 (scientific ambiguity, different units, numerical precision).

Next, we wanted to see how question type dependence is influenced by which specific system (search engine, number of queries and documents per question) is used. Figure 4 provides an overview of the same systems discussed in Section 3.2.

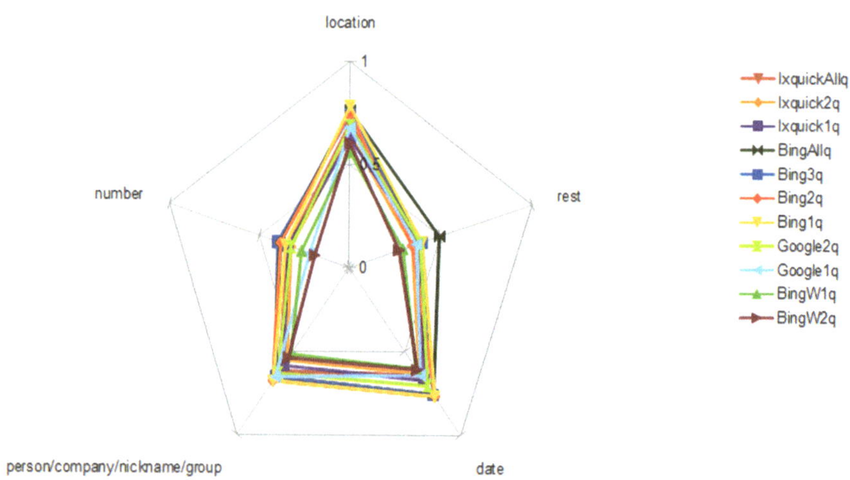

Fig. 4: netdiagram diversity.

3.4 System Combination

Looking at the previous line of experiments, it is obvious that systems behave differently for different answer types. For example, the Google1q system performs decent on names, poor on numbers, and average on the other categories. In contrast, Ixquick2q performs poor on names and much better on numbers. This observation inspires the use of system combination to stimulate synergetic behavior.

When combining systems, we had to make sure that the n-best list of answers produced by the systems are identical. If an answer included in the n-best list of System A was missing in the n-best list of System B, it was included in the latter with a confidence of 0. Then, the individual n-best lists were merged multiplying the individual confidence scores with system-dependent weights and summing the results up across systems. This resulted in final n-best list the highest confidence result was considered winner. The system weights are greater or equal to zero and need to sum up to one. An example for tuning the weight of the system Ixquick20q when combined with Ixquick200q is shown in Figure 5. The peak is found at a weight of 0.3 where the system performs at 57.4% compared to the individual systems (52.6% and 56.1%). Minimum p-value is 0.14, that is, the effect is moderately statistically significant.

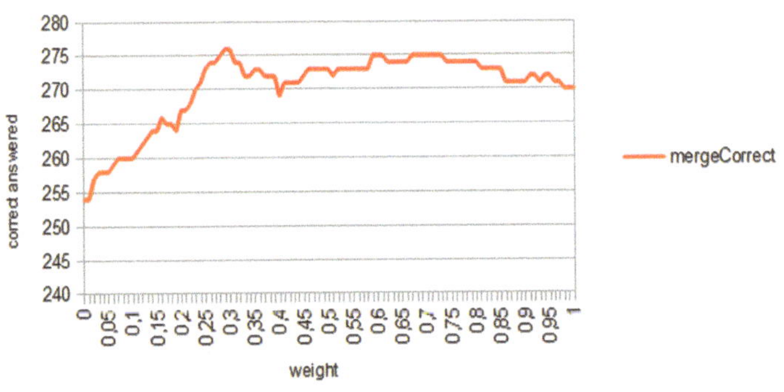

Fig. 5: system combination w. Bing20q & Bing200q

4 Implementation in Halef

To demonstrate OpenEphyra's capabilities, we integrated it into the spoken dialog system Halef [3], an open-source industry-standard-compliant, distributed system. The system can be called using the free-of-charge number

```
+1-206-203-5276 Ext 2000
```

As the current system is limited to rule-based (JSGF) grammars, the set of possible questions that can be asked is limited to the ones encoded in the grammar, e.g.

Who discovered gravity?
When was Albert Einstein born?
Who invented the automobile?

5 Conclusion and Future Work

This paper features a number of contributions:

- We discussed the issue of outdated test sets in question answering, analyzed reasons, and possible ways of remedy.
- We presented the results of a number of tuning experiments with the open-source question answering system OpenEphyra.
- We desribed how we changed OpenEphyra's interface to external knowledge bases from a commercial search API to the direct connection to the web search engines Google, Bing, and Ixquick.
- We showed how by extensive parameter tuning and system combination, the new web search interface can perform en par with the original implementation based on a commercial search API.

In the future, we aim at addressing underperforming answer types (in particular numbers and, to some extent, names) and breaking the rest group down into multiple sub-groups each of which can be tackled independently [13].

Acknowledgements

We would like to thank all the members of Spoken Dialog Systems Research Center at DHBW Stuttgart. Furthermore, we wholeheartedly appreciate the continuous support of Patrick Proba and Monika Goldstein.

Raimo Bakis, Jiří Havelka, and Jan Cuřín

Meta-Learning for Fast Dialog System Habituation to New Users

Abstract: When a new user comes on board a mobile dialog app, the system needs to quickly learn this user's preferences in order to provide customized service. Learning speed depends on miscellaneous parameters such as probability smoothing parameters. When a system already has a large number of other users, then we show how meta-learning algorithms can leverage that corpus to find parameter values which improve the learning speed on new users.

1 Motivation

You arrive at the office in the morning and say to your new assistant:

> *"Get me Joe on the phone, please."*
> The assistant looks at your address book: *"Is that Joe Smith or Joe Brown?"*
> You say, *"Brown."*
> Next morning, again, *"I need to talk to Joe."*
> Assistant: *"Brown, right? Not Smith?"*
> You: *"Yes, Brown."*
> The next time you want Joe, the assistant probably doesn't even ask, maybe just confirms: *"OK, calling Joe Brown."*

A psychologist might call the assistant's changing responses "habituation," a clearly useful characteristic of human behavior, necessary for establishing rapport between individuals. But that assistant might not be a person – it could be a software app. That app, too, must habituate quickly. To achieve this, we show that it needs to draw on its experiences with other users it has encountered before you. It learns *your preferences* by observing *you*, but it must first learn *how to learn* by observing a sample of *other users*.

Rapport, of course, is a two-way street. While the assistant learns to anticipate your needs, you come to know the assistant's mind. For example, after calling Joe

Raimo Bakis: IBM T. J. Watson Research Center, 1101 Kitchawan Rd, Yorktown Heights, NY 10598, USA, e-mail: bakis@us.ibm.com
Jiří Havelka, Jan Cuřín: IBM Prague Research Lab, V Parku 2294/4, Prague, 148 00, Czech Republic, e-mail: {havelka, jan_curin}@cz.ibm.com

Brown a few times, you know you must now say the full name if you want Joe Smith, or the assistant would dial the wrong Joe.

For efficient communication with an electronic assistant, that device, too, needs to behave in a sufficiently human-like manner to enable the user to develop a sense of mutual understanding. This is another reason for software to mimic human traits like habituation.

Finally, a mystery challenge: you often talk to Joe Brown in the morning, but not in the afternoon, and now it is late in the day, and you are calling a man named Joe. Is it likely to be the same person?

2 Related Work

In recent years, there has been a great amount of research into statistical dialog systems. Partially Observable Markov Decision Processes (POMDPs) offer a general framework for implementing a statistical dialog manager (DM) module, consisting of a belief estimation sub-module (modeling the DM's beliefs about the user's state) and an action policy sub-module (specifying what action to take in a belief state). POMDPs represent a principled and theoretically well-grounded approach to implementing statistical spoken dialog system, see Young et al. (2013) for a survey of the current state of art in the development of POMDP-based spoken dialog systems.

Even though a POMDP can be used as a theoretical model for implementing the whole DM in a dialog system, doing so is a challenging and highly non-trivial task, requiring a model of the domain and a source of data for optimizing hidden parameters of the model. From a mundane point of view, replacing an existing rule-based dialog management module in a spoken dialog system might be infeasible due to practical reasons and costs.

In this work we present an approach that allows us to take a step in the direction of incremental introduction of statistical modules into dialog systems, be it into existing or newly developed dialog applications. We concentrate on modeling beliefs over sub-parts of the full dialog state that can be used by hand-crafted or automatically learned policies for sub-dialogs. Such "local" eliefs can be exposed to a dialog app developer in a simple and non-intrusive way, allowing him or her to take advantage of them and potentially greatly simplify the implementation of dialog flow (policy), while at the same time improving the quality of the app, for example by shortening average task completion time.

Discovering users' preferences and using them appropriately is essential for providing personalized user experience, see e.g. Jannach and Kreutler (2005). Our approach could be described as non-intrusive preference elicitation where information gathered across users is used for tuning shared parameters that allow for quickly adapting a particular user's model, thus mitigating the so-called *new user problem* (poor models for new users).

We show that it is possible to estimate beliefs in an unknown and changing environment (such as, for example, a user's preferred contacts in his/her changing contact list) in a uniform manner, allowing the use of dialog policies common to all users. We believe that this addresses environments different from those generally employed in POMDP-based dialog systems, which tend to be stable and shared by all users, such as bus scheduling (Black et al. 2011) or tourist information domains (Young et al. 2010).

3 Outline

In the next section, we introduce meta-learning using a small pilot experiment, based on actual mobile-user dialogs, in which the system learns to efficiently estimate probabilities of user intents in a scenario similar to the hypothetical one introduced above. Subsequently, in section 5, we view this experiment in a larger context and discuss extended applications of similar principles. There we also address the question of the mysterious contact you are calling late in the day, which we mentioned at the end of section 1.

4 Meta-Learning Introduced via a Pilot Experiment

We address our motivating problem from section 1: Given a "short list" of contacts, our goal in this experiment is to reduce the DM's remaining uncertainty – measured in terms of entropy – about the user's intent to as low a value as possible.

4.1 Data Sample

From a sample consisting of logs of 2,790 dialog sessions on mobile devices, we extracted all those where a user asked to make a phone call but the dialog manager (DM) was unable to determine a unique phone number on the basis of the user's initial request. The DM then presented a list of choices to the user, both in a GUI display and by Text-To-Speech. We selected only those dialogs in which the user subsequently made a unique selection from that list and where the DM then dialed that number. We found 134 such dialogs – these constitute the sample for the following discussion. We will assume that the telephone number which the DM finally dialed in each of these dialogs represented user's true intent for that dialog.

In all the dialogs in this sample, a rule-based component of the DM had already determined that the user's intent was to make a phone call, and had furthermore narrowed down the possible intended contacts to a short list. In this experiment a statistical algorithm then attaches a probability or "belief" to each candidate in the short list. Our discussion applies to a generic statistical algorithm; in section 4.3 we present a concrete instance of such an algorithm and show how we can tune it using a data sample.

4.2 Formulation as Optimization Problem

We formulate the above task as an optimization problem, which allows us to find optimal values of parameters of a statistical algorithm of our interest using standard numerical optimization methods.

Let K be the number of dialogs; in the present study, $K = 134$. Let I_k be the length of the contact short list in the k-th dialog, where $k = 1, 2, \ldots, K$.

Now consider the contact who appears in position i in that list. Let $n_{k,i}$ be the total number of times this contact has already been called by the same person who is the user in the k-th dialog. Furthermore, let i_k be the index of that user's true intent in the short list in this, the k-th dialog.

The statistical algorithm calculates a belief (subjective probability) $b_k(i)$ for the i-th contact in the short list in the k-th dialog. These values are normalized so that

$$\sum_{i=1}^{I_k} b_k(i) = 1 \quad \forall\, k \in \{1, 2, \ldots, K\}.$$

Now let $p_k(i)$ designate the true probability of the i-th candidate in the k-th dialog. Because we assume we know the user's actual intent with certainty, we have $p_k(i_k) = 1$ and $p_k(j) = 0$ for all $j \in \{1, 2, \ldots, I_k\}$ such that $j \neq i_k$. In that case, the entropy of the true probabilities is zero, $H(p_k) = 0$, and hence the cross-entropy $H(p_k, b_k)$ becomes equal to the Kullback-Leibler divergence $D_{\mathrm{KL}}(p_k \| b_k)$ of the beliefs b_k from the true probabilities p_k, which in turn becomes equal to $-\log b_k(i_k)$.

We now define our objective function as the average value of that cross-entropy:

$$F = -\frac{1}{K} \sum_{k=1}^{K} \log b_k(i_k). \tag{1}$$

Clearly, if the beliefs were always correct, so that

$$b_k(i_k) = 1 \quad \forall\, k \in \{1, 2, \ldots, K\},$$

then F would achieve its minimum value, $F = 0$. In practice, we will generally not reach that value, but our aim is to minimize F.

Whenever we can evaluate the gradient of an objective function with respect to its parameters, we can try to find optimal values of the parameters using standard numerical optimization methods, such as gradient descent or L-BFGS (Liu and Nocedal 1989). In our case, we would tune any relevant parameters of a statistical algorithm implementing belief estimation of our choice.

4.3 A Concrete Algorithm – Laplace Smoothing

When a statistical algorithm calculates $b_k(i)$, it is, in principle, allowed to use all information in existence at the time of that computation. In section 5 we discuss the use of several kinds of data.

In this pilot study, however, we restricted that computation to use only the counts $n_{k,i}$ and one additional parameter α. We denote this by writing

$$b_k(i \mid n_{k,1}, \ldots, n_{k,I_k}, \alpha).$$

When, for some k, all counts are large, that is

$$n_{k,i} \gg 1 \quad \forall\, i \in \{1, 2, \ldots, I_k\},$$

then it might be reasonable to use the maximum-likelihood estimator for all $i = 1, 2, \ldots, I_k$

$$b_k(i \mid n_{k,1}, \ldots, n_{k,I_k}, \alpha) = \frac{n_{k,i}}{\sum_{j=1}^{I_k} n_{k,j}}. \tag{2}$$

But we are interested in hitting the ground running, so to speak. Our probability estimates must be as accurate as possible even when $n_{k,i} = 0$ for some k and i. In those cases Equation (2) would clearly be wrong, because it would set $b_k(i) = 0$ whenever $n_{k,i} = 0$. This would imply that the DM knew with certainty that the user was not calling candidate i in dialog k. For example, if by the time dialog k started, the user had called Joe Brown once but had never yet called Joe Smith, and if that user did now want to call Joe Smith, then the DM would have to assume that it mis-heard, or that the user mis-spoke, because according to Equation (2) the probability of calling Joe Smith in dialog k would be zero – an exact mathematical zero with no wiggle-room.

But if Equation (2) is wrong, then what is the true probability of calling somebody new, somebody this user has never called before with this app? That question – how to calculate the probability of a hitherto unseen event – has vexed philosophers for centuries. What is the probability that the sun will not rise tomorrow? Or the probability of a black swan? Fortunately, in our application, we don't have to philosophize or speculate, we can determine the answer empirically through meta-learning.

A frequently used method for getting rid of the zero probabilities for unseen events is probability smoothing. Perhaps the simplest and oldest way of doing this is additive smoothing, also known as Laplace smoothing (Manning et al. 2008). This method simply adds a "smoothing constant" α to each observed count n. The probability estimate for candidate i in the k-th dialog becomes then for all $i = 1, 2, \ldots, I_k$

$$b_k(i \mid n_{k,1}, \ldots, n_{k,I_k}, \alpha) = \frac{n_{k,i} + \alpha}{\sum_{j=1}^{I_k} (n_{k,j} + \alpha)}. \tag{3}$$

Under certain theoretical assumptions, the correct value for the smoothing constant would be $\alpha = 1$. In realistic situations, however, this is generally not the optimal value. Imagine, for example, a population of users who called all contacts in their phone books equally often. Then it is easy to see that the DM should ignore the observed counts – these would be just statistical noise. Optimal smoothing, in that case, would be achieved when $\alpha \to \infty$, which would yield equal probabilities for all candidates in the short list. At the other extreme, consider a population where every user only ever called one person, regardless of how many contacts they had in their phone book. In that case $\alpha = 0$ would be the correct value. Real populations of users fall somewhere in between. That is why it is necessary to determine α empirically. Within any given population of users, of course, there is also variation of behaviors, but we can at least determine the best value for the population average.

Figure 1 shows how the average cross-entropy defined by Equation (1) depends on α in our sample. For this population of users we found the optimal value to be $\alpha = 0.205$. This is significantly lower than the sometimes-quoted theoretical value of 1, but is definitely larger than zero, which vindicates that meta-learning can indeed improve belief estimation for new users using data from a population of other users.

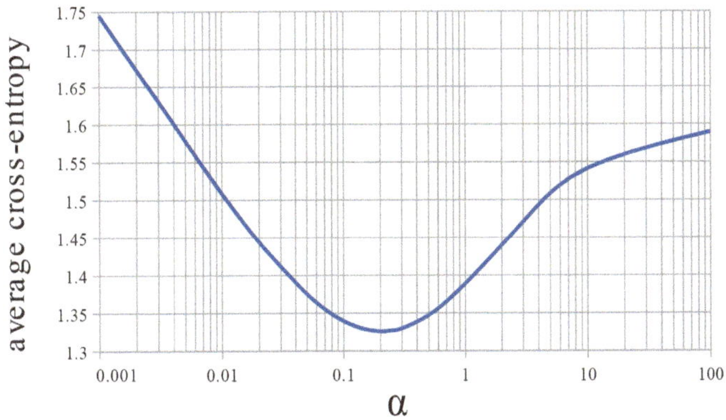

Fig. 1: Average cross-entropy depending on smoothing constant α

5 Future Work

The beliefs calculated from Equation (3) are the same in the morning as in the evening. To accommodate the hypothetical example in the introduction where you might call Joe Brown more often in the morning, we could let the values $n_{k,i}$ in Equation (3) be functions of the time of day. Let φ be the time-of-day phase of the proposed new call: 0 at midnight, 2π next midnight. Then we might write

$$n_{k,i}(\varphi) = \sum_{m=0}^{M} (\alpha_{m,k,i} \cos m\varphi + \beta_{m,k,i} \sin m\varphi) \tag{4}$$

or we might prefer a log-linear model where

$$n_{k,i}(\varphi) = \exp\left(\sum_{m=0}^{M} (\alpha_{m,k,i} \cos m\varphi + \beta_{m,k,i} \sin m\varphi) \right). \tag{5}$$

The α and β coefficients would be computed by some algorithm from not only the total number of previous calls to the same contact by the same caller, but also from the times of these calls. Space does not permit a detailed discussion here, but algorithms related to Fourier analysis are obvious candidates. These algorithms would involve parameters affecting smoothing across candidates as well as smoothing across time. All such parameters would be optimized empirically in ways analogous to that illustrated in the previous section.

The predicted beliefs could be allowed to depend on many other variables as well, in addition to the time of day. Obviously, a user's preferences change as time passes, which could be modeled using a decay parameter (a function of elapsed time) applied to individual past observations. The user's geographic location at the time of the call may predict which pizza shop gets the order. The probability of calling a taxi service may depend on the weather, etc.

When the beliefs are calculated as functions of quantities for which we have good intuitive models, then we can design special-purpose functions such as those in Equations (3)–(5). When this is not possible, then more general families of parametric functions may be appropriate, such as artificial neural nets, see for example Bishop (2006), Rumelhart et al. (1986).

6 Summary

Cognitive behavior on the part of a mobile device – behavior which appears thoughtful and intelligent to the user – requires, among other things, that the system anticipate users' wishes and intents as quickly as possible. How fast and how accurately it learns a new user's preferences depends on parameter settings which, in turn, must be learned from a sample of other users from the same population. In this report, we illustrated this principle by means of a simple probability-smoothing example, and we sketched extensions of the algorithm for handling more complex patterns.

Bibliography

Bishop, C.M. (2006) *Pattern Recognition and Machine Learning*. Springer

Black, A.W., Burger, S., Conkie, A., Hastie, H., Keizer, S., Lemon, O., Merigaud, N., Parent, G., Schubiner, G., Thomson, B., et al. (2011) Spoken dialog challenge 2010: Comparison of live and control test results. In: *Proceedings of the SIGDIAL 2011 Conference*. pp. 2–7. Association for Computational Linguistics

Jannach, D., Kreutler, G. (2005) Personalized User Preference Elicitation for e-Services. In: Cheung, W., Hsu, J. (eds.), *Proceedings of the 2005 IEEE International Conference on e-Technology, e-Commerce and e-Service*. pp. 604–611. IEEE

Liu, D.C., Nocedal, J. (1989) On the limited memory BFGS method for large scale optimization. *Mathematical Programming* 45(1–3):503–528

Manning, C.D., Raghavan, P., Schütze, H. (2008) *Introduction to Information Retrieval*. Cambridge University Press, Cambridge, UK

Rumelhart, D.E., McClelland, J.L., et al. (1986) *Parallel Distributed Processing: Volume 1: Foundations*. MIT Press, Cambridge, MA

Young, S., Gašić, M., Thomson, B., Williams, J. (2013) POMDP-Based Statistical Spoken Dialog Systems: A Review. *Proceedings of the IEEE* 101(5):1160–1179

Young, S., Gašić, M., Keizer, S., Mairesse, F., Schatzmann, J., Thomson, B., Yu, K. (2010) The Hidden Information State Model: a practical framework for POMDP-based spoken dialogue management. *Computer Speech & Language* 24(2):150–174

Martin Mory, Patrick Lange, Tarek Mehrez, and
David Suendermann-Oeft

Evaluation of Freely Available Speech Synthesis Voices for Halef

Abstract: We recently equipped the open-source spoken dialog system (SDS) Halef with the speech synthesizer Festival which supports both unit selection and HMM-based voices. Inspired by the most recent Blizzard Challenge, the largest international speech synthesis competition, we sought to find which of the freely available voices in Festival and those of the strongest competitor Mary are promising candidates for operational use in Halef. After conducting a subjective evaluation involving 36 participants, we found that Festival was clearly outperformed by Mary and that unit selection voices performed en par, if not better, than HMM-based ones.

1 Introduction

Spoken dialog systems (SDSs) developed in academic environments often substantially differ from those in industrial environments [1, 2] with respect to the data models (statistical vs. rule-based), the use cases (demonstration vs. productive usage), the underlying protocols, APIs, and file types (self-developed vs. standardized ones) and many more criteria.

To bridge the gap between both worlds, we have developed the SDS Halef (*H*elp *A*ssistant *L*anguage-*E*nabled and *F*ree) [3], being entirely based on open-source components, mainly written in Java. Unlike other academic SDSs, Halef's architecture is distributed, similarly to industrial SDSs; see Figure 1. In addition, Halef adheres to the following industrial standards:

- JSGF (Java Speech Grammar Format) as rule-based speech recognition grammar format [4],

Martin Mory: DHBW Stuttgart, Stuttgart, Germany and Hewlett-Packard, Böblingen, Germany, e-mail: martin.mory@hp.com
Patrick Lange: DHBW Stuttgart, Stuttgart, Germany and Linguwerk, Dresden, Germany and Staffordshire University, Stafford, UK, e-mail: patrick.lange@linguwerk.de
Tarek Mehrez: DHBW Stuttgart, Stuttgart, Germany and University of Stuttgart, Stuttgart, Germany and German University in Cairo, Cairo, Egypt, e-mail: tarekmmehrez@gmail.com
David Suendermann-Oeft: DHBW Stuttgart, Stuttgart, Germany, e-mail: david@suendermann.com

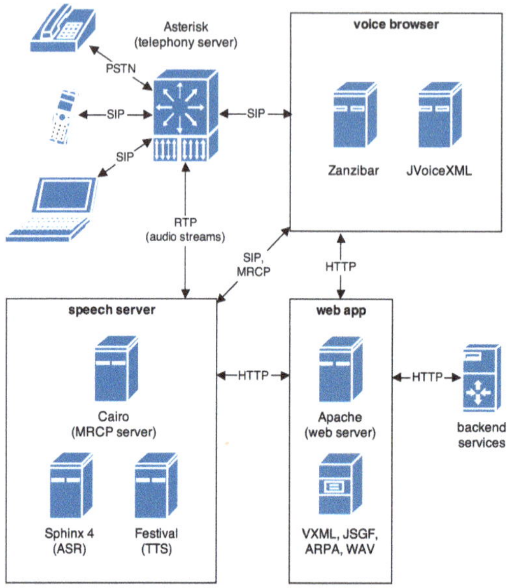

Fig. 1: High-level architecture of Halef.

- MRCP (Media Resource Control Protocol) [5] for exchange coordination of voice browser, speech recognition and synthesis components,
- RTP (Real-time Transport Protocol) [6] for audio streaming,
- SIP (Session Initiation Protocol) [7] for telephony, and
- VoiceXML [8] for dialog specification.

Simplifying the usual process for pursuing speech-driven applications, Halef provides a framework for implementing such applications by supplying the speech recognition and synthesis resources, leaving the dialog's logic as the only variable, to be controlled by the user.

The question answering system demonstrated in [3] is the first application being realized with Halef. The current version of Halef is limited to the English language, however, we are working on two additional applications which will be based on German (an information system for Stuttgart's public transportation authority and an intoxication checkup).

The quality of the speech synthesis component is crucial for the usability of a spoken dialog system. Commercially, this quality controls how appealing the system is for the ordinary user. Originally, Halef was based on the obsolete speech synthesizer FreeTTS [9] whose active development was terminated years ago and which caused user complaints about the synthesis quality. Our first step to improve quality was to enable our platform to use modern HMM-based voices. We did so by integrating the TTS system Festival [10] developed at the Centre for Speech Technology Research of the University of Edinburgh. This pushed Halef few steps forward towards a high-performing research-based SDS.

The decision to use Festival as the new TTS system was based on results of the last Blizzard Challenge [11], the world's most comprehensive speech synthesis evaluation. The results indicated that the HTS [12] voices of Festival were rated better than those of the other free TTS systems.

As it turned out, a number of HTS voices being accessible in the Festival online demo, and likely the ones achieving the best performance in the Blizzard challenge, are not publicly available. That is why the results and conclusions from the Blizzard Challenge cannot necessarily be applied to our setting. Therefore, we decided to conduct an own study to compare all those voices which are freely available to the public.

As indicated by the Blizzard challenge, another TTS system of prime quality is Mary [13], developed by the DFKI (German Research Center for Artificial Intelligence). We included Mary's English voices in our study as well. Table 1 provides an overview about the voices we compared in the evaluation. This set of voices covers the different synthesis principles of unit selection as well as statistical parametric synthesis.

Tab. 1: Overview of the compared voices.

voice	system	technique
cmu-bdl-hsmm	Mary	HSMM
cmu-rms-hsmm	Mary	HSMM
cmu-slt-hsmm	Mary	HSMM
cmu_us_awb_cg	Festival	Clustergen [14]
cmu_us_clb_arctic_clunits	Festival	cluster unit selection
cmu_us_rms_cg	Festival	Clustergen
cmu_us_slt_arctic_hts	Festival	HMM
dfki-obadiah-hsmm	Mary	HSMM
dfki-obadiah	Mary	unit selection
dfki-poppy-hsmm	Mary	HSMM
dfki-poppy	Mary	unit selection
dfki-prudence-hsmm	Mary	HSMM
dfki-prudence	Mary	unit selection
dfki-spike-hsmm	Mary	HSMM
dfki-spike	Mary	unit selection
kal_diphone	Festival	diphone unit selection

2 Experimental Setup

In order to get a reliable rating of the 16 freely available TTS voices, subjects had to assess samples of each of these voices, letting them rate their absolute quality on a six-level Likert scale. The levels from which to chose were 'utterly bad' (1), 'poor' (2), 'okay' (3), 'fine' (4), 'good' (5) and 'excellent' (6). Each voice had to be rated using exactly one of the named levels and independently of other voices. The sentence "This is some example speech to evaluate the quality of available Festival and Mary voices" was used as the reference text being synthesized by all the 16 voices. The test audience consisted of persons who are familiar with speech processing and laypersons as well, ensuring that both specialist and non-specialist perceptions are taken into consideration.

As main measure to compare voices in this study, we calculated the mean opinion score (MOS) of the quality score described above. To test the significance of conclusions, we are using the Welch's t test since we are comparing independent samples expected to be Gaussian distributed. We assumed a significance level of 5%.

3 Results

Table 2 shows the results of the survey. For each voice and each level, the count of subjects who rated the voice with that score is given. The right column provides the MOS by which the table is sorted. The total number of participants was 36.

4 Discussion

4.1 Ranking

The voice dfki-spike achieved the highest MOS (4.39), but the Welch test was unable to confirm that it is significantly better than the voices dfki-obadiah and dfki-spike-hsmm. However, dfki-spike was found to be significantly better than the voice cmu-bdl-hsmm and all the ones with worse MOS. Therefore, the first three voices can be interpreted as a cluster of the best rated voices in the study. The voices dfki-poppy, dfki-prudence, cmu-rms-hsmm, and dfki-prudence-hsmm are very close as well and form the second best-rated cluster. The third cluster is formed by the voices cmu_us_slt_arctic_hts, cmu_us_rms_cg, cmu-slt-hsmm,

Tab. 2: Evaluation results, sorted by MOS.

voice	(1)	(2)	(3)	(4)	(5)	(6)	MOS
dfki-spike	0	4	5	9	9	9	4.39
dfki-obadiah	1	4	6	10	11	4	4.06
dfki-spike-hsmm	0	5	7	11	9	4	4.00
cmu-bdl-hsmm	0	2	16	8	9	1	3.75
dfki-poppy	3	7	7	6	11	2	3.58
dfki-prudence	3	8	8	5	7	5	3.56
cmu-rms-hsmm	1	7	9	12	6	1	3.50
dfki-prudence-hsmm	1	7	13	8	5	2	3.42
cmu_us_slt_arctic_hts	1	6	15	8	6	0	3.33
cmu_us_rms_cg	1	7	16	6	5	1	3.28
cmu-slt-hsmm	2	11	9	9	3	2	3.17
dfki-poppy-hsmm	4	12	9	3	7	1	3.00
dfki-obadiah-hsmm	2	14	10	5	4	1	2.94
cmu_us_awb_cg	12	18	6	0	0	0	1.83
kal_diphone	20	13	1	0	2	0	1.64
cmu_us_clb_arctic_clunits	27	9	0	0	0	0	1.25

dfki-poppy-hsmm, and dfki-obadiah-hsmm, all of which do not show significant mean differences in the Welch test. The fourth cluster, consisting of the three worst voices of the study is concluded by the Arctic cluster unit selection voice.

4.2 Unit Selection vs. HSMM

The study included a number of voice pairs with the same speaker but different synthesis techniques. For each of the unit selection voices dfki-spike, dfki-obadiah, dfki-poppy, and dfki-prudence, there are HSMM-based counterparts. For dfki-poppy and dfki-obadiah, the unit selection version was rated significantly better than their HSMM-based counterparts. The two other pairs show the same tendence without being significantly different. Apparently, the natural sound of the unit selection voices was perceived to be more important for the test audience than the artifacts caused by the unit concatenation. This is an unexpected result, underlining that the evaluation of voice quality is very subjective and showing that unit selection can fulfill expectations in practical usage.

4.3 Festival vs. Mary

Our experiment shows that the results of the Blizzard Challenge [11] cannot be applied to our setting. Taking into consideration that the best rated Festival voice in our study is significantly worse than multiple Mary voices and furthermore that the three worst voices are Festival voices, it is unambiguous that Mary outperformed Festival based on the set of freely available voices. This is further evidenced by the MOS of Mary (3.58) versus that of Festival (2.27). The Festival unit selection voices clearly sound artificial, the accentuation is incorrect or missing at all and several sound distortions inhibit the understanding of the synthesized speech. Opposed to that, the Mary unit selection voices show only very few distortions and seem to have been recorded at higher sample rates which greatly improves the quality of the synthesis.

5 Conclusions

We presented results of a study comparing the synthesis quality of freely available voices of the speech synthesizers Festival and Mary. It turned out that Mary clearly outperforms Festival with the three best rated voices dfki-spike, dfki-obadiah, and dfki-spike-hsmm, because these voices are ahead of the Festival voices in the field in terms of better emphasis, less sound distortions and more natural sounding. Surprisingly, two of the three winners are unit-selection-based voices.

6 Future Work

Having identified the three best out of all voices available to us, we decided to equip Halef with the speech synthesizer Mary. To make sure that the present study's results are applicable to spoken dialog systems in operation, we will conduct a second evaluation round where subjects will rate full conversions with Halef rather than isolated recordings.

Acknowledgements

We thank all participants of our subjective evaluation.

Bibliography

[1] Pieraccini, R., Huerta, J.: Where Do We Go from Here? Research and Commercial Spoken Dialog Systems. In: Proc. of the SIGdial, Lisbon, Portugal (2005)

[2] Schmitt, A., Scholz, M., Minker, W., Liscombe, J., Suendermann, D.: Is It Possible to Predict Task Completion in Automated Troubleshooters? In: Proc. of the Interspeech, Makuhari, Japan (2010)

[3] Mehrez, T., Abdelkawy, A., Heikal, Y., Lange, P., Nabil, H., Suendermann-Oeft, D.: Who Discovered the Electron Neutrino? A Telephony-Based Distributed Open-Source Standard-Compliant Spoken Dialog System for Question Answering. In: Proc. of the GSCL, Darmstadt, Germany (2013)

[4] Hunt, A.: JSpeech Grammar Format. W3C Note. http://www.w3.org/TR/2000/NOTE-jsgf-20000605 (2000)

[5] Burnett, D., Shanmugham, S.: Media Resource Control Protocol Version 2 (MRCPv2). http://tools.ietf.org/html/rfc6787 (2012)

[6] Schulzrinne, H., Casner, S., Frederick, R., Jacobsen, V.: RTP: A Transport Protocol for Real-Time Applications. Technical Report RFC 3550, IETF (2003)

[7] Johnston, A.: SIP: Understanding the Session Initiation Protocol. Artech House, Norwood, USA (2004)

[8] Oshry, M., Auburn, R., Baggia, P., Bodell, M., Burke, D., Burnett, D., Candell, E., Carter, J., McGlashan, S., Lee, A., Porter, B., Rehor, K.: VoiceXML 2.1. W3C Recommendation. http://www.w3.org/TR/2007/REC-voicexml21-20070619 (2004)

[9] Walker, W., Lamere, P., Kwok, P.: FreeTTS: A Performance Case Study. Technical report, Sun Microsystems, Santa Clara, USA (2002)

[10] Taylor, P., Black, A., Caley, R.: The Architecture of the Festival Speech Synthesis System. In: Proc. of the ESCA Workshop on Speech Synthesis, Jenolan Caves, Australia (1998)

[11] Charfuelan, M., Pammi, S., Steiner, I.: MARY TTS Unit Selection and HMM-Based Voices for the Blizzard Challenge 2013. In: Proc. of the Blizzard Challenge, Barcelona, Spain (2013)

[12] Yamagishi, J., Zen, H., Toda, T., Tokuda, K.: Speaker-Independent HMM-based Speech Synthesis System - HTS-2007 System for the Blizzard Challenge 2007. In: Proc. of the ICASSP, Nevada, USA (2008)

[13] Schroeder, M., Trouvain, J.: The German Text-to-Speech Synthesis System MARY: A Tool for Research, Development and Teaching. International Journal of Speech Technology **6**(4) (2001)

[14] Black, A.W.: Clustergen: a statistical parametric synthesizer using trajectory modeling. In: INTERSPEECH, ISCA (2006)

Michael Zock and Dan Cristea

You Shall Find the Target Via its Companion Words: Specifications of a Navigational Tool to Help Authors to Overcome the Tip-Of-The-Tongue Problem

Abstract: The ability to retrieve 'words' is a prerequisite for speaking or writing. While this seems trivial as we succeed most of the time, it is not simple at all and it can become quite annoying if ever we fail or require too much time. The resulting silence puts pressure on the speaker's and the listener's mind, as it disrupts the fluency of encoding (planning what to say) and decoding (interpretation of the linguistic form). When lacking information for a given 'word' (for example, its form), we tend to reach for a dictionary. While this works generally quite well for the language receiver, this is not always the case for the language producer. This may be due to a number of reasons like input presentation (underspecification), organisation of the lexicon, etc. We present here the roadmap of a lexical resource whose task it to help authors to find the word they are looking for. More precisely, we present here a framework for building a tool to support word access. To reach our goal several problems need be solved: 'search space reduction', 'clustering the words retrieved in response to some input (available information)' and 'labeling the clusters' to ease navigation. Before starting to build this navigational tool, we define a set of criteria that need to be satisfied by the resources to be used. Next we discuss some of them to see whether they comply with respect to our goal. While being preliminary work, this is clearly a necessary step for building the tool we have in mind.

Michael Zock: Aix-Marseille Université, CNRS, LIF UMR 7279, 13000, Marseille, France, e-mail: michael.zock@lif.univ-mrs.fr
Dan Cristea: "Alexandru Ioan Cuza" University of Iaşi and Institute of Computer Science, Romanian Academy, e-mail: dcristea@info.uaic.ro

1 The Problem: How to Find the Word that is Eluding You

One of the most vexing problems in speaking or writing is that one knows a given word, yet one fails to access it when needed. Suppose, you were looking for a word expressing the following ideas: 'superior dark coffee made of beans from Arabia', but could not retrieve the intended form 'mocha'. What will you do in a case like this? You know the meaning, you know how and when to use the word, and you even know its form, since you've used it some time ago, yet you simply cannot access it at the very moment of speaking or writing? Since dictionaries generally contain the target word, they are probably our best ally to help us find the form we are looking for. This being said, storage does not guarantee access. The very fact that a dictionary contains a word does not guarantee at all that we will also be able to find or locate it (Zock & Schwab, 2013; Tulving & Pearlstone, 1966).

Dictionary users typically pursue one of two goals (Humble, 2001): as *decoders* (reading, listening), they are generally interested in the meanings of a specific word, while as *encoders* (speakers, writer) they wish to find the form expressing an idea or a concept. This latter task is our goal. While most dictionaries satisfy the reader's needs, they do not always live up to the authors' expectations, helping them to find the elusive word.[1] There are various reasons for this. Some of them are related to the input:

(a) *Input specification*: What information should one provide to look up a specific word? This is not a trivial issue, even for quite common words, say, 'car', 'apple' or 'elephant. Should the input be a single word, a set of words ('huge, gray, Africa', in the case of elephant), a more general term (category, animal), or a textual fragment (context) from which only the target is missing? Con-

---

1 To be fair, one must admit though that great efforts have been made to improve the situation. In fact, there are quite a few onomasiological dictionaries. For example, Roget's Thesaurus (Roget, 1852), analogical dictionaries (Boissière, 1862, Robert et al., 1993), Longman's Language Activator (Summers, 1993) various network-based dictionaries: WordNet (Fellbaum,1998; Miller et al., 1990), MindNet (Richardson et al., 1998), and Pathfinder (Schvaneveldt, 1989). There are also various collocation dictionaries (BBI, OECD), reverse dictionaries (Kahn, 1989; Edmonds, 1999) and OneLook which combines a dictionary, WordNet, and an encyclopedia, Wikipedia (http://onelook.com/reverse-dictionary.shtml). A lot of progress has been made over the last few years, yet more can be done especially with respect to indexing (the organization of the data) and navigation. Given the possibilities modern computers offer with respect to storage and access, computational lexicography should probably jettison the distinctions between lexicon, encyclopedia, and thesaurus and unify them into a single resource.

cerning this last point, see for example the shared task of SemEval, devoted to lexical substitution (McCarthy & Navigli, 2009).

(b) *Synonymy* or *term-equivalence*: suppose your target (chopstick) were defined as 'instrument used for eating', yet you used the query term 'tool' instead of 'instrument'.

(c) *Ambiguity*: how does the machine (lexical resource) 'know' which of the various senses you have in mind ('mouse/device' vs. 'mouse/rodent')?[2]

Others are related to the output produced in response to some input:

(a) *Number of outputs*: since entries (query terms) can be very broad, i.e. linguistically underspecified (suppose you were to use 'animal' in the hope to find 'elephant'), the number of hits or outputs can be huge. Hence we must organize them. But size can become critical even if one uses other search strategies, as we will do. We try to find the target via its associated term. Since in both cases the list of outputs is huge, we must address this problem, and we believe that the answer lies in clustering or organization.

(b) *Cluster names*: outputs must not only be grouped, but the groups need names, as otherwise the user does not know in what direction to go, that is, in what bag to look for the target.

Search failure (failing to find) is called *dysnomia* or *Tip of the Tongue-problem* (Brown & McNeill, 1996)[3] if the searched objects are words. Yet, this kind of problem occurs not only in communication, but also in other activities of everyday life. Being basically a search problem, it is likely to occur whenever we look for something that exists in real world (objects) or our mind: dates, phone numbers, past events, peoples' names, or 'you-just-name-it'.

As one can see, we are concerned here with the problem of words, or rather, how to find them in the place where they are stored: lexical resource (dictionary or brain). We will present here some ideas of how to develop a tool in order to help authors (speaker/writer) to find the word they are looking for. While there are various search scenarios, we will restrict ourselves here only to cases where the searched terms exists in the base. Our approach is based on psychological

2 Note that outputs can also be polysemous, but ambiguity is not really a problem here, as all the dictionary user wants is to find a given word form.

3 The tip-of-the-tongue phenomenon (http://en.wikipedia.org/wiki/Tip_of_the_tongue) is characterized by the fact that the author (speaker/writer) has only partial access to the word s/he is looking for. The typically lacking parts are phonological (syllables, phonemes). Since all information except this last one seems to be available, and since this is the one preceding articulation, we say: the word is stuck on the *tip of the tongue* (TOT, or TOT-problem).

findings concerning the mental lexicon (Levelt, 1989).[4] Hence we draw on notions such as association (Deese, 1965), associative network (Schvaneveldt, 1989) or neighborhood (Vitevitch, 2008). We also take into account notions such as *storage* (representation and organization), *access* of information (Roelofs, 1992; Levelt et al., 1999), observed *search strategies* (Thumb, 2004) and typical *navigational behavior* (Atkins, 1998). Our goal is to develop a method allowing people to access words, no matter how incomplete their conceptual input may be. To this end we try to build an index, i.e. a semantic map allowing users to find a word via navigation.

2 Search Strategies Function of Variable Cognitive States

Search is always based on knowledge. Depending on the knowledge available at the onset one will perform a specific kind of search. Put differently, there are different information needs as there are different search strategies.

There are at least three things that authors typically know when looking for a specific word: its *meaning* (definition) or at least part of it (this is the most frequent situation), its *lexical relations* (hyponymy, synonymy, antonymy, etc.), and the collocational or encyclopedic relations it entertains with other words (Paris-city, Paris-French capital, etc.). Hence there are several ways to access a word (see Figure 1): via its meaning (concepts, meaning fragments), via syntagmatic links (thesaurus- or encyclopedic relations), via its form (rhymes), via lexical relations, via syntactic patterns (search in a corpus), and, of course, via another language (translation).

Suppose you were looking for a (word)form expressing the following ideas 'spring, typically found in Iceland, discharging intermittently hot water and steam'. The corresponding word, 'geyser' or 'geysir' can be recovered in various ways:

4 While paper dictionaries store word forms (lemma) and meanings next to each other, this type of information is distributed across various layers in the mental lexicon. This may lead to certain word access problems. Information distribution is supported by many empirical findings like *speech errors* (Fromkin, 1980), studies in *aphasia* (Dell *et al.*, 1997), experiments on *priming* (Meyer & Schvaneveldt, 1971) or the *tip of the tongue phenomenon* (Brown & McNeill, 1996). For computer simulations see (Levelt et al., 1999; Dell, 1986).

1. directly based on its *meaning* (this is the golden, i.e. normal route). Note that google does quite well for this kind of input. We will also be able to do that, but we can do a lot more, namely, reveal the target via its associates.
2. by searching in the corresponding semantic field: 'hotsprings' or 'natural fountains encountered in Iceland', etc;
3. by relying on encyclopaedic information (co-occurrences, associations): hotspring, Iceland, eruption, Yellowstone;
4. by considering similar *sounding* words (Kaiser geyser);
5. via a *lexical relation* (synonym, hypernym) like "fountain", hotspring;
6. via a syntactic pattern (co-occurrence): 'spring typically found in Iceland;'
7. *translation* equivalent (**間歇噴泉** geyser);

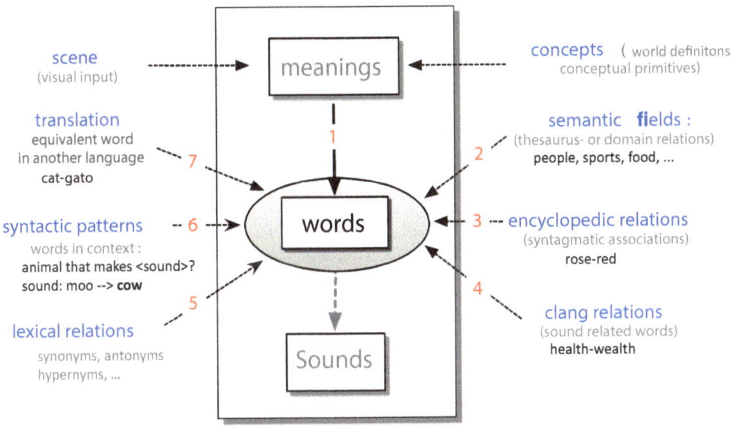

Fig. 1: Seven routes or methods for accessing words.[5]

We will consider here only one strategy, the use associations (mostly, encyclopaedic relations). Note that, people being in the TOT-state clearly know more than that. Psychologists who have studied this phenomenon (Brown & McNeill, 1966; Brown, 2012; Díaz et al. 2014; Schwartz and Metcalfe, 2011) have found that their subjects had access not only to *meanings* (the word's definition), but also to information concerning *grammar* (for example, gender, see Vigliocco et al. 1997) and *lexical form*: sound, morphology (part of speech). While all this information

5 This feature of the *mental lexicon* (ML) is very important, as in case of failure of one method, one can always resort to another.

could be used to constrain the search space, – the ideal dictionary being multiply indexed, – we will deal here only with semantically related words (associations, collocations in the large sense of the word). Before discussing how such a dictionary could be built and used, let us consider a possible search scenario.

We start from the assumption that in our mind, all words are connected, the mental lexicon (brain) being a network. This being so, anything can be reached from anywhere. The user enters the graph by providing whatever comes to his mind (source-word), following the links until he has reached the target. As has been shown (Motter et al. 2002), our mental lexicon has small-world properties: very few steps are needed to get from the source-word to the target word. Another assumption we make is the following: when looking for a word, people tend to start from a close neighbour, which implies that users have some meta-knowledge containing the topology of the network (or the structure of their mental lexicon): what are the nodes, how are they linked to their neighbours, and what are more or less direct neighbours? For example, we know that 'black' is related to 'white', and that both words are fairly close, at least a lot closer than, say, 'black' and 'flower'.

Search can be viewed as a dialogue. The user provides as input the words that a concept he wishes to express evokes, and the system displays then all (directly) connected words. If this list contains the target search stops, otherwise it will continue. The user chooses a word of the list, or keys in an entirely different word. The first part described is the simplest case: the target is a direct neighbour. The second addresses the problem of indirect associations, the distance being bigger than 1.

3 Architecture and Roadmap

As mentioned already, when experiencing word access problems we expect help from dictionaries, hoping to find the elusive term there. Unfortunately, so far there is still not yet a satisfying resource allowing authors (people being in the 'production mode': speakers/writers) to find easily and most of the time the resisting word. While WordNet or Roget's Thesaurus are helpful in some cases, more often than one might think, they are not. This is a problem we would like to overcome. Figure 2 displays in a nutshell our approach, word access being viewed (basically) as a two-step process: two for the user, and two for the resource builder. The task is basically finding a specific item (target word) within the lexicon. Put differently, the task is to reduce the entire set (all words contained in the lexicon) to one, the target. Since it is out of question to search in the entire lexicon, we suggest to reduce the search space in several steps, basically two.

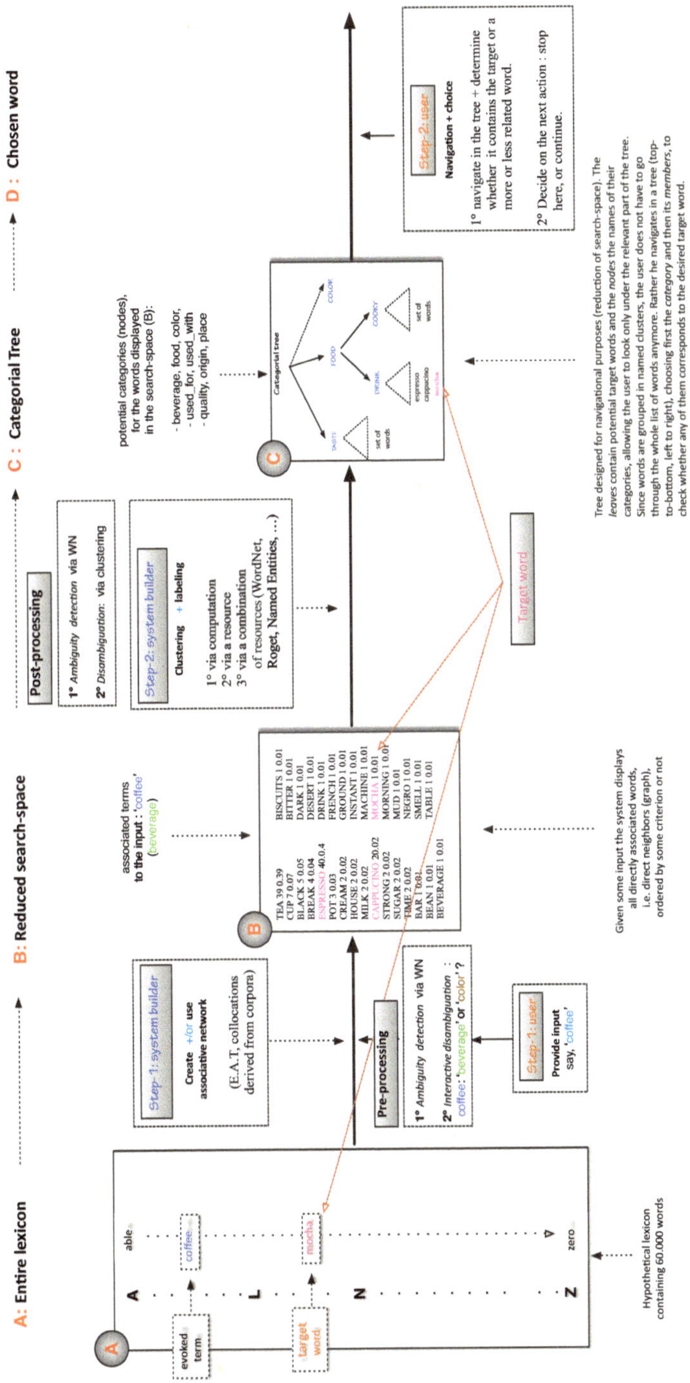

Fig. 2: Lexical access as a two-step process.

The goal of the first step is to reduce the initial space (about 60.000 in the case of EAT) to a substantially smaller set. EAT[6] is an association thesaurus which generates all directly related words to some input (the word given by the user, word coming to his mind when trying to find the form of a given concept).

The goal of the second step is to reduce further the search space. Since the list of words directly associated to the input is still quite huge (easily 150–500 words), we suggest to cluster and label the terms to build a categorial tree. This allows the user to search only in the relevant part of the categorial tree, rather than in the entire list, leaving him finally with a fairly small number of words. If all goes well, he will find the target in this tree, otherwise, he will have to iterate, either starting from an entire new word or choosing one contained in the selected cluster.

Note, that in order to display the right search space, i.e. set of words within which search takes place (step-1), we must have already well understood the input – [mouse (rodent) vs. mouse(device)] – as otherwise our set may contain many (if not mostly) inadequate candidates: 'cat/cheese' instead of 'computer/screen' or vice versa. Note also, that the ideal resource should allow us to solve both problems: allow for navigation in an associative network while presenting the potential candidates in meaningfully named clusters (categorial tree). Given the complexity of the task at hand, we will certainly not try to start building such a resource. Rather we will try to 'discover' the best one among the existing resources. Hence, we will propose here below a methodology to evaluate the advantages and shortcomings of a number of notorious resources in correlation with our problem at hand.

4 Formalisation of Resources and Discussion

To evaluate the adequacy of the various resources to be considered when building our tool requires some criteria, which we call *properties*. Let's note that some of them are binary (yes/no) hence likely to act as constraints, while others are gradual, they could be valued. Here is an initial set of such properties:

1. Representation: undirected graph (constraint). In this graph vertexes are words[7] (possibly ambiguous), or word senses[8] (therefore, unambiguous), and

6 http://www.eat.rl.ac.uk

7 Whenever we use the term 'word' we imply not only single terms but also 'collocations' or 'multiword expressions', that is, a sequence of words expressing meaning.

8 A word sense should be understood as an indexed word. For a word, there are as many indexes as the word form has senses.

edges are undirected relations. The reason why directionality of edges can be ignored is that access should be bi-directional. At this stage we will also ignore possible names of relations (labels), in order to remain as general as possible, and to be able to consider the greatest number of resources.

2. Completeness: gives an estimation concerning the size of the lexicon (valued). If we aim at retrieving any word of a language, the resource should include all its words. This feature can only be evaluated in fuzzy terms, because no dictionary is or can be complete, be it for reasons related to proper nouns (named entities), newly coined terms (neologisms), etc.

3. Connectivity: connected graph (constraint). This means that all words should be connected. The graph shall not contain any isolated nodes. This is an obvious property if our goal is to allow reaching a target from any input word.

4. Density: the average number of connections each node has with its neighbour nodes (valued). A small number would not be good. Let us take an extreme case. Imagine a lexicon in which words are ordered alphabetically, and the only connexions from a word are towards its immediate neighbours, the previous and the next word in the dictionary. This arrangement defeats nearly all chances to find the target, as, in the worst case, one would have to traverse the whole lexicon in order to get from the source- to the target-word. At the other extreme are graphs with an extremely large number of connections. This is also undesirable. Imagine a totally connected lexicon, one where each word is linked to all the others (complete graph). This would not work neither. Even though the target word is always included in the list as direct neighbour, being surrounded by a great number of irrelevant words, it cannot be found in due time.

5. Features: each node of the graph should be characterised by a set of features, to be used later on for filtering and clustering (valued). Hence, once a word is spotted (step-1) it should spread activation to all its neighbours, possibly filtered according to some shared properties. This would yield clusters which still need to be named. Both operations are feature-based.

We present now a comparative analysis of some well-known resources along the lines just described.

WordNet (WN)

1. *Representation*: un-directed graph (passed). Vertexes here are synsets rather than words, and edges are relations (hypernymy/hyponymy/antonymy/etc.), ignoring their intrinsic directionality. Synsets are not only lists of equivalent words.

Actually all words (or literals) being part of a synset have attached with them an index showing their word sense. For example the word “cop” is an element of 3 separate synsets, cop#1, cop#2 and cop#3, meaning respectively: policeman, to steal, take into custody. This being so it is easy to translate all synsets into their respective word senses.

2. *Completeness*: close to 100% for common words, but very low for proper nouns (at least for the Princeton WN version);

3. *Connectivity*: failed, because the lexicon is split in 4 isolated graphs: nouns, verbs, adjectives and adverbs (with very few cross-POS links: play, tennis). Nouns and verbs are connected internally, because of the hierarchy, but there is no guarantee for adjectives and adverbs;

4. *Density*: rather low, as the links correspond to the small set of relations dealt with in WN (about a dozen). Hence, every word sense displays only very few links;

5. *Features*: POS, LEMMA[9], SENSE, but also sets of linked words as given by relations (even if they form also edges of the graph): HYPERNYMY-SET, HYPONYMY-SET, ANTONYMY-SET, etc.

Let us note that we could have a variant of this resource with nodes representing word forms (actually lemmas) rather than word senses. To this end one would clash all nodes representing different senses into a single node, merge the edges and operate a transformation on features. As such, the representation condition will pass too, the completeness will not change, and the connectivity will now be higher, because polysemous words will collect links from all its previous sense nodes. The POS and LEMMA features will not change (being identical in all previous nodes representing word senses), but SENSE will combine with each of the HYPERNYMY-SET, etc. (because the semantic relations characterise word senses and not words).

Extended WN (Ext-WN)

1. *Representation*: un-directed graph (passed, as above). In this resource, unlike in WN, the elements of a gloss are semantically disambiguated. Hence a gloss is likely to yield a rich set of links towards all the respective synsets;

9 Or sequence of lemmas, in case of collocations. Since, a node is here a word sense, its lemma should be considered as a feature.

2. *Completeness*: same as for WN;

3. *Connectivity*: perhaps passed, because the links added by the glosses allow to cross the POS barrier;

4. *Density*: higher than WN (apart from the traditional WN links, links generated by the glosses are added here);

5. *Features*: POS, LEMMA, SENSE, but also HYPERNYMY-SET, HYPONYMY-SET, ANTONYMY-SET, etc. (as the sets of words connected by the respective relations) and GLOSS-SET (word senses issued in the annotated glosses).

A similar variant as the one mentioned for WN, where nodes are words and not word senses, can be thought for Ext-WN.

Edinburg Association Thesaurus (EAT)
1. *Representation*: undirected graph with nodes being words and edges associations (if considered in both directions) (passed);

2. *Completeness*: The EAT network contains 23,219 vertices (words) and 325,624 arcs (stimulus-response pairs). For what concerns us here, EAT contains about 56,000 words, which is much less than WN (perhaps not passed);

3. *Connectivity*: yet to be proved (the more associations are displayed for each source word, the higher the chance to obtain connectivity, but we are not aware of any empirical proof that a path can be drawn between any two words in this resource);

4. *Density*: medium;

5. *Features*: ASSOCIATION-SET (an explicit, direct link), POS (implicit, can be deduced). Here SENSE is not marked. Note that two other points that may count are the number of stimulus words and the number of responses produced to some input.

Note that there are quite a few other attempts to build word association lists. For example, the Free Association Thesaurus[10] is probably the largest resource of this

10 http://web.usf.edu/FreeAssociation/Intro.html

kind for American English. It produces 750,000 responses to 5,019 stimulus words. The goal of the '*small world of words*' project[11] is to build a multi-lingual map of the human lexicon. At present it contains more than five million responses for Dutch and more than a million for English. Given their size, these resources are quite likely to pass the test of completeness.

An unstructured language corpus (a representative collection of sentences of a language)
Let's note that word occurrences in a corpus are distinct from words of the language (lexemes, or title words, as they appear in a dictionary or a thesaurus). In a corpus, the context could be exploited to build the connectivity, the context of the word occurrence w^{occ} being, for instance, all word occurrences (excepting for itself) belonging to the sentence to which w^{occ} belongs to. But, making a graph in which nodes are word occurrences yields a collection of small, disconnected graphs, ruining all chances for navigation. This is clearly an example of a bad representation for a resource like the one we need. To repair this, we need to connect the words belonging to different sentences to each other. One way of achieving this would be to integrate all occurrences of the same word into a single node.

1. *Representation*: an un-directed graph where nodes are considered to be words; Hence a word w has an edge towards another word w_j if the corpus contains a sentence with w and w_j appearing together (passed).

2. *Completeness*: complete if the corpus is large enough (passed);

3. *Connectivity*: passed, if the corpus is large enough;

4. *Density*: very high, perhaps even unmanageable for high frequency words, yet rather small for the rest. In accordance to Zipf's law, the density of the edges follows a power law distribution.

5. *Features*: POS, LEMMA, SENTENCE_ID X WORDS (this last feature is the vector product of the set of sentence IDs and a subset of all words; thus word occurrences are modelled by storing for each word/node a pair containing the ID of the sentence it belongs to and the list of all the other content words in that sentence).

A variant of this resource can be a graph in which nodes represent word senses.

11 http://fac.ppw.kuleuven.be/lep/concat/

Roget's Thesaurus

1. *Representation*: the graph could be thought of as a collection of isolated word entries (passed).

2. *Completeness*: complete[12] (passed);

3. *Connectivity*: failed, as words are isolated;

4. *Density*: zero;

5. Features: CLASS, SECTION, SUB-SECTION, HEAD-GROUP, HEAD, POS, PARAGRAPH, SEMICOLON, LEMMA.

The representation, proposed above, does not support navigation. However, Roget's Thesaurus has the merit to present a rich set of features attached to words and it is a good candidate to combinations.

Other resources could be taken into consideration: Wikipedia, when considered in combination with its hierarchy of categories, DBpedia, BabelNet, ConceptNet, etc.

The comparison of the resources along the various dimensions gives us an idea concerning their relative adequacy with respect to our goal. Note that we did not take into account the last property, called 'features'. It seems that WN fails in terms of connectivity and density. Ext-WN escapes the connectivity problem, but its density is probably still too low to allow for a significant expansion in Step-1. EAT does not pass the completeness property and seems weak with respect to connectivity. The density of a corpus is likely to be extremely unbalanced, displaying either too many or too few links, depending on the word. Roget's Thesaurus fails both in terms of connectivity and density.

To overcome the weaknesses of individual resources with respect to the TOT problem, combinations of resources could be considered. Obviously a combination of two resources is allowed if and only if the representation constrains with respect to nodes and edges are identical. This means that nodes should represent either words or word senses in both resources, in which case the combination will be formed by simply merging edges and combining features of identical nodes.

For instance, we could combine Ext-WN with EAT to overcome the density problem of Ext-WN. To this end we could combine the links generated by glosses

12 Digitised by Jarmasz (2003), based on the 1987 version of Roget, published by Pearson Education.

with EAT's associations. This would increase considerably the number of candidates at the end of Step-1. WN as well as Ext-WN seem to lend themselves well for the clustering operation referred to as Step-2. In both cases one could use the hypernymy links, at least for nouns: for instance, a group of words having the same hypernym (closest common ancester) could be clustered together, while having already naturally a name, given by any member of the hypernym synset, or the whole set altogether[13].

5 Outlook and Conclusion

To summarize, we were dealing here with *word access* by people being in the **production mode**. Word finding is viewed as an interactive, fundamentally cognitive process. It is *interactive* as it involves two agents who cooperate (human/computer), and it is *cognitive* as it is based on knowledge. Since this latter is incomplete for both of them, they cooperate: neither of them alone can point to the target word, but working together they can. Having complementary knowledge they can help each other to find the elusive word.

How this can be accomplished precisely remains to be clarified in further work. Meanwhile we have sketched a formal representation of linguistic resources, on which a clustering and naming general strategy could be applied. While so far no single resource seems to be adequate to offer a satisfying solution, combining the right ones should yield a tool, allowing users to overcome the TOT-problem. While our ultimate goal is to help authors to *find* what they can't *recall* based on whatever they can *remember*, at present we can offer only preliminary solutions. Clearly, a lot of work lies ahead of us.

Acknowledgements

Part of the work of the second author was done under the project *The Computational Representative Corpus of Contemporary Romanian Language*, a project of the Romanian Academy.

13 Please note that the distance of the various elements with respect to a common hypernym may be quite variable, hence cluster names may vary considerably in terms of abstraction.

Bibliography

[1] Atkins, S. (ed.), (1998). Using Dictionaries. Studies of Dictionary Use by Language Learners and Translators. Tübingen: Max Niemeyer Verlag.

[2] Boissière, P. (1862). Dictionnaire analogique de la langue française: répertoire complet des mots par les idées et des idées par les mots. Paris. Auguste Boyer

[3] Brown, A. S. (2012). The tip of the tongue state. New York: Psychology Press.

[4] Brown, R. & Mc Neill, D. (1966). The tip of the tongue phenomenon. Journal of Verbal Learning and Verbal Behavior, 5:325–337

[5] Deese, J. (1965). The structure of associations in language and thought. Johns Hopkins Press. Baltimore

[6] Dell, G., Schwartz, M., Martin N., Saffran E. & Gagnon D. (1997). Lexical access in aphasic and nonaphasic speakers. Psychol Rev. 1997 Oct; 104(4):801–38.

[7] Dell, G. (1986). A spreading-activation theory of retrieval in sentence production. *Psychological Review*, 93:283–321.

[8] Díaz, F., Lindín, M., Galdo-Álvarez, S. & Buján, A. (2014). Neurofunctional Correlates of the Tip-of-the-Tongue State. In Schwartz, B. W. & Brown, A. S. (2014). Tip of the tongue states and related phenomena. Cambridge University Press.

[9] Edmonds, D. (ed.), (1999). The Oxford Reverse Dictionary, *Oxford University Press*, Oxford, 1999.

[10] Fellbaum, C. (1998). WordNet: An Electronic Lexical Database and some of its Applications. MIT Press.

[11] Fromkin, V. (ed.), (1980). *Errors in linguistic performance: Slips of the tongue, ear, pen, and hand*. San Francisco: Academic Press.

[12] Humble, P. (2001). Dictionaries and Language Learners, Haag and Herchen.

[13] Jarmasz, M. (2003). Roget's Thesaurus as a Lexical Resource for Natural Language Processing. PhD thesis, Ottawa-Carleton Institute for Computer Science

[14] Kahn, J. (1989). Reader's Digest Reverse Dictionary, *Reader's Digest*, London

[15] Levelt, W. (1989). Speaking: From intention to articulation. Cambridge, MA: MIT Press.

[16] Levelt, W., Roelofs, A. & Meyer, A.S. (1999). A theory of lexical access in speech production. Behavioral and Brain Sciences, 22:1–75.

[17] McCarthy, D. & Navigli, R. (2009). The English lexical substitution task. Language resources and evaluation, 43(2):139–159.

[18] Meyer, D.E. and Schvaneveldt, R.W. (1971). Facilitation in recognizing pairs of words: Evidence of a dependence between retrieval operations. Journal of Experimental Psychology 90:227–234.

[19] Miller, G.A. (ed.), (1990): WordNet: An On-Line Lexical Database. International Journal of Lexicography, 3(4):235–244.

[20] Motter, A. E., A. P. S. de Moura, Y.-C. Lai, and P. Dasgupta. (2002). Topology of the conceptual network of language. Physical Review E, 65(6), (4):107–117.

[21] Richardson, S., Dolan, W. & Vanderwende, L. (1998). Mindnet: Acquiring and structuring semantic information from text. In: ACL-COLING'98. Montréal: 1098–1102.

[22] Robert, P., Rey A. & Rey-Debove, J. (1993). Dictionnaire alphabetique et analogique de la Langue Française. Le Robert, Paris.

[23] Roelofs, A. (1992). A spreading-activation theory of lemma retrieval in speaking. In Levelt, W. (ed.), Special issue on the lexicon, Cognition, 42:107–142.

[24] Roget, P. (1852). Thesaurus of English Words and Phrases. Longman, London.
[25] Schvaneveldt, R. (ed.), (1989). Pathfinder Associative Networks: studies in knowledge organization. Ablex. Norwood, New Jersey, US.
[26] Schwartz, B., & Metcalfe, J. (2011). Tip-of-the-tongue (TOT) states: retrieval, behavior, and experience *Memory & Cognition, 39* (5):737–749
[27] Summers, D. (1993). Language Activator: the world's first production dictionary. Longman, London.
[28] Thumb, J. (2004). Dictionary Look-up Strategies and the Bilingualised Learner's Dictionary. A Think-aloud Study. Tübingen: Max Niemeyer Verlag.
[29] Tulving, E., & Pearlstone, Z. (1966). Availability versus accessibility of information in memory for words. Journal of Verbal Learning and Verbal Behavior, 5:381–391
[30] Vigliocco, G., Antonini, T. & Garrett, M. F. (1997). *Grammatical gender is on the tip of Italian tongues*. Psychological Science, 8:314–317.
[31] Vitevitch, M. (2008). What can graph theory tell us about word learning and lexical retrieval? Journal of Speech, Language, and Hearing Research, 51:408–422.
[32] Zock, M. & Schwab, D (2013) L'index, une ressource vitale pour guider les auteurs à trouver le mot bloqué sur le bout de la langue. In Gala, N. et M. Zock (eds.), Ressources lexicales: construction et utilisation. Lingvisticae Investigationes, John Benjamins, Amsterdam, The Netherlands, pp. 313–354.

Ali M. Naderi, Horacio Rodríguez, and Jordi Turmo

Topic Modeling for Entity Linking Using Keyphrase

Abstract: This paper proposes an Entity Linking system that applies a topic modeling ranking. We apply a novel approach in order to provide new relevant elements to the model. These elements are keyphrases related to the queries and gathered from a huge Wikipedia-based knowledge resource.

1 Introduction

Recently, the needs of world knowledge for Artificial Intelligence applications are highly increasing. As a part of world knowledge, Knowledge Bases (KB) are appropriate for both human and machine readability, involved to keep and categorize entities and their relations. The KB profits by improving the ability of obtaining more amount of discriminative information in a shorter range of time than discovering through all unstructured resources. But the high cost of manual elicitation to create KB forces toward automatic acquisition from text. This requires two main abilities. 1) extracting relevant information of mentioned entities including attributes and relations between them (Slot Filling), and 2) linking these entities with entries in the ontology (Entity Linking–EL). This paper focuses on the latter.

EL is the task of linking an entity mention occurring in a document (henceforth, background document) to a unique entry within a reference KB, e.g. when seeing the text "American politician Chuck Hagel", if the involved KB is Wikipedia (WP), the entity mention "Chuck Hagel" should be linked to the WP entry http://en.wikipedia.org/wiki/Chuck_Hagel. Assigning the correct reference of entities such as persons, organizations, and locations is highly challenging since one entity can be referred to by several mentions (synonymy), as well as same mention may be used to depict distinct entities (ambiguity). For instance, "George Yardley" might refer to either the Scottish former footballer, or the American basketball player (ambiguity), who is also known by nicknames such as "Yardbird" or shortly "Bird" (synonymy). The ambiguity can be more challenging, e.g. in the sentence "they have Big Country in this NBA match.", the surface form "Big Country" is

Ali M. Naderi, Horacio Rodríguez, Jordi Turmo: TALP Research Center, UPC, Spain, e-mail: {anaderi,horacio,turmo}@lsi.upc.edu

referring to "Bryant Reeves", the NBA professional basketball player. In Discussion Fora (DF) such as blogs, etc. the texts might contain grammatical irregularities which make the EL even harder, e.g. consider the sentence "James Hatfield is working with Kirk Hammett". The surface form "James Hatfield" can be referred to the American author, but the correct grammatical form of "Hatfield" is "Hetfield" referring to the main songwriter and co-founder of heavy metal band Metallica. These synonymy and ambiguity challenges make it difficult for natural language processors to realize the correct reference of entity mentions in the text. In addition, as further challenges faced to the EL, an entity can be mentioned in a text by its partial names (rather than its full name), acronyms or other types of name variation.

This paper proposes an Entity Linking system that applies a topic modeling ranking to face to the ambiguity problem. We apply a novel approach in order to provide elements of the model by taking advantage of keyphrases gathered from a huge WP-based knowledge resource.

2 Literature Review

The recent works on EL in its contemporary history are inspired from the older history of Word Sense Disambiguation (WSD) where this challenge firstly arised. Many studies achieved on WSD are quite relevant to EL. Disambiguation methods in the state of the art can be classified into supervised methods, unsupervised methods and knowledge-based methods [17].

Supervised Disambiguation (SD). The first category applies machine-learning techniques for inferring a classifier from training (manually annotated) data sets to classify new examples. Researcher proposed different methods for SD. A *Decision List* [22] is a SD method containing a set of rules (if-then-else) to classify the samples. In continue, [10] used learning decision lists for *Attribute Efficient Learning*. [13] introduced another SD method *Decision Tree* that has a tree-like structure of decisions and their possible consequences. *C4.5* [19], a common algorithm of learning decision trees was outperformed by other supervised methods [16]. [9] studied on the *Naive Bayes* classifier. This classifier is a supervised method based on the Bayes' theorem and is a member of simple probabilistic classifiers. The model is based on the computing the conditional probability of each class membership depending on a set of features. [16] demonstrated good performance of this classifier compared with other supervised methods. [14] introduced *Neural Networks* that is a computational model inspired by central nervous system of organisms. The model is presented as a system of interconnected neurons. Although [24] showed an appropriate performance by this model but the experiment was achieved in a small size of data. However, the dependency to large amount of training data is a major drawback [17]. Recently, different combination of supervised approaches are proposed. The combination methods are highly interesting since they can cover the weakness of each stand-alone SD methods [17].

Unsupervised Disambiguation (UD). The underlying hypothesis of UD is that, each word is correlated with its neighboring context. Co-located words generate a cluster tending to a same sense or topic. No labeled training data set or any machine-readable resources (e.g. dictionary, ontology, thesauri, etc.) are applied for this approach [17]. *Context Clustering* [23] is a UD method by which each occurrence of a target word in a corpus is indicated as a context vector. The vectors are then gathered in clusters, each indicating a sense of target word. A drawback of this method is that, a large amount of un-labeled training data is required. [12] studied on *Word Clustering* a UD method based on clustering the words which are semantically similar. Later on, [18] proposed a word clustering approach called *clustering by committee* (CBC). [25] described another UD method *Co-occurrence Graphs* assuming that co-occurrence words and their relations generate a co-occurrence graph. In this graph, the vertices are co-occurrences and the edges are the relations between co-occurrences.

Knowledge-based Disambiguation (KD). The goal of this approach is to apply knowledge resources (such as dictionaries, thesauri, ontologies, collocations, etc.) for disambiguation [2, 3, 5, 11, 15]. Although, these methods have lower performance compared with supervised techniques, but they have a wider coverage [17].

Recently, some collective efforts are done to research in this field in form of challenging competitions. The advantage of such competitions is that, the performance of systems are more comparable since all participants assess their systems in a same testbed including same resources and training and evaluation corpus. To this end, *Knowledge Base Population* (KBP) EL track at *Text Analysis Conference* (TAC) [1] is the most important challenging competition being subject of significant study since 2009. The task is annually organized by which many teams present their proposed systems.

1 The TAC is organized and sponsored by the U.S. National Institute of Standards and Technology (NIST) and the U.S. Department of Defense.

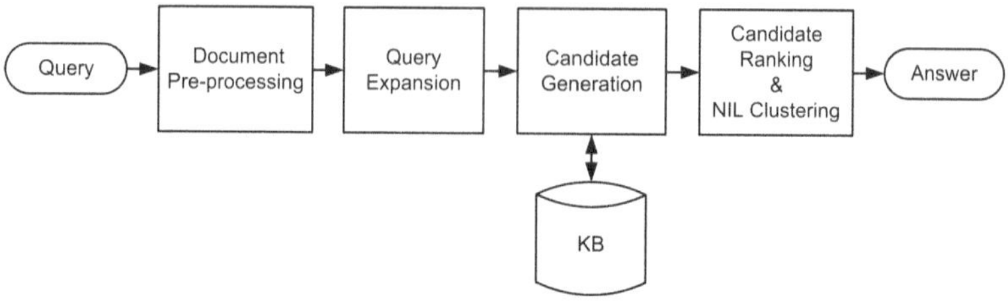

Fig. 1: General architecture of the EL systems.

3 Methodology and Contribution

The method proposed in this paper follows the typical architecture in the state of the art (Figure 1). Briefly, given a query, consisting of an entity mention and a background document, the system preprocesses the background document (Document Pre-processing step). Then, the background document is expanded integrating more related and discriminative information corresponding to each query in order to facilitate finding the correct reference of each query mention in the KB (Query Expansion step). Subsequently, those KB nodes which can be potential candidates to be the correct entity are selected (Candidate Generation step). Finally, the candidates are ranked in a top-down hierarchy and the candidate having the highest order is selected. Furthermore, all queries belonging to the same Not-In-KB (NIL) entity are clustered together assigning the same NIL id (Candidate Ranking and NIL clustering step). The final task (Candidate Ranking and NIL Clustering) is the most challenging and highly crucial among steps above. In order to rank candidates, we apply topic modeling. As a contribution, we take advantage of keyphrases to enrich the background document in the Query Expansion step in order to improve the performance of the system in ranking candidates.

Details of each step are provided next.

3.1 Document Pre-Processing

Initially, the background document must be converted to a standard structure to be used by other components. To this objective, the system pre-processes the document in the following way.

Document Partitioning and Text Cleaning. This component separates the textual and non-textual parts of the document. Then, the further steps are only applied over the textual part.

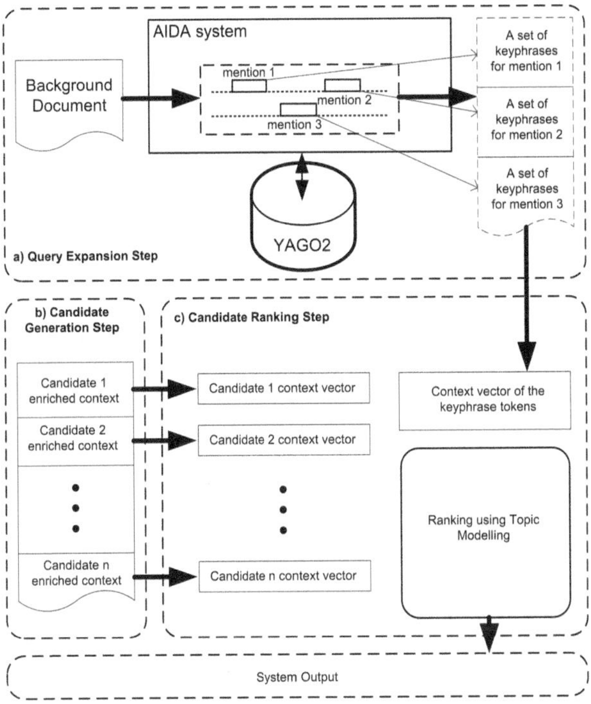

Fig. 2: Detailed architecture of applying keyphrases for ranking candidates.

In addition, each document might contain several HTML tags and noise (e.g. in Web documents) which are removed by the system.

Sentence Breaking and Text Normalization. This module operates on the context of documents as following:

- *Sentence Breaking.* The documents are splitted by discovering sentence boundaries.
- *Capitalization.* Initial letters of words occurring in titles and all letters of acronyms are capitalized.
- *Soft Mention (SM).* Entity mentions represented with abbreviations are expanded, e.g. "Tech. Univ. of Texas", is replaced with "Technical University of Texas". To this end, a dictionary-based mapping are applied.

Tab. 1: List of tables in YAGO2.

dictionary	entity_counts
entity_ids	entity_inlinks
entity_keyphrases	entity_keywords
entity_lsh_signatures_2000	entity_rank
keyphrase_counts	keyword_counts
meta	word_expansion
word_ids	

3.2 Query Expansion

In most queries, query name might be ambiguous, or background document contains poor and sparse information about the query. In these cases, query expansion can reduce the ambiguity of query name and enrich the content of documents through finding name variants of the query name, integrating more discriminative information, and tagging meta data to the content of documents.

For doing so, we apply the following techniques:

Query Classification. Query type recognition helps to filter out those KB entities with type different to the query type. Our system classifies queries into 3 entity types: PER (e.g. "George Washington"), ORG (e.g. "Microsoft") and GPE (GeoPolitical Entity, e.g. "Heidelberg city"). We proceed under the assumption that a longer mention (e.g. "George Washington") tends to be less ambiguous than a shorter one (e.g. "Washington") and, thus, the type of the longest query mention tends to be the correct one. The query classification is performed in three steps. First, we use the Illinois Named Entity Recognizer and Classifier (NERC) [20] to tag the types of named entity mentions occurring in the background document. Second, we find the set of mentions in the background document referring to the query. More concretely, we take mention m_1 defined by the query offsets within the background document (e.g. "Bush") and find the set of mentions that include m_1 (e.g. "Bush", "G. Bush", "George W. Bush"). Finally, we select the longest mention from the resulting set of mentions and take its type as the query type.

Background Document Enrichment. This task includes two subsequent steps applied to the background document: a) *mention disambiguation*, and b) *keyphrase exploitation* for each mention. As explained in Section 3.4, we apply *Vector Space Model* (VSM) for ranking candidates. As VSM components are extracted from the background document of each query, we need as most disambiguated entities as possible. For doing so, AIDA system [8] is applied. AIDA is useful for entity de-

Tab. 2: The entity id for the entity "Bill_Gates" (Table 2a), the information of a sample keyphrase for this entity (Table 2b), and the associated keyphrase name (with the length of 1 token) with the keyphrase id 18098 (Table 2c).

(a)

entity	id
Bill_Gates	134536

(b)

entity	keyphrase	keyphrase_ tokens	keyphrase_ token_weights	source	count	weight
134536	18098	{18098}	0.0001	linkAnchor	56	0.009

(c)

word	id
IBM	18098

tection and disambiguation. Given an unstructured text, it maps entity mentions onto entities registered in YAGO2 [7], a huge semantic KB derived from WP, WordNet [4], and Geonames[2]. YAGO2 contains more than 10 million entities and around 120 million facts about these entities. Using AIDA, the system disambiguates as much as possible mentions in the content of background document. Table 1 shows the list of tables in YAGO2 containing the structured information about entities. Each entity in YAGO2 contains several types of information, including weighted keyphrases. A keyphrase which can be used to disambiguate entities, is contextual information extracted by YAGO authors from link anchor, in-link, title and WP category sources of the relevant entity page. For instance, the keyphrase information related to a named entity mention e.g. "Bill Gates" occurring in the background document is respectively obtained from YAGO2 by following SQL commands:

i. `select * from entity_ids where entity='Bill_Gates';`
ii. `select * from entity_keyphrases where entity=134536;`
iii. `select * from word_ids where id=18098;`

The Tables 2a, 2b, and 2c show the commands output respectively. We appended a new component to AIDA system to automatically gather the necessary keyphrases of each entity mention from YAGO2 (performing the SQL commands mentioned above) in order to use these keyphrases in EL task (Figure 2a).

2 www.geonames.org

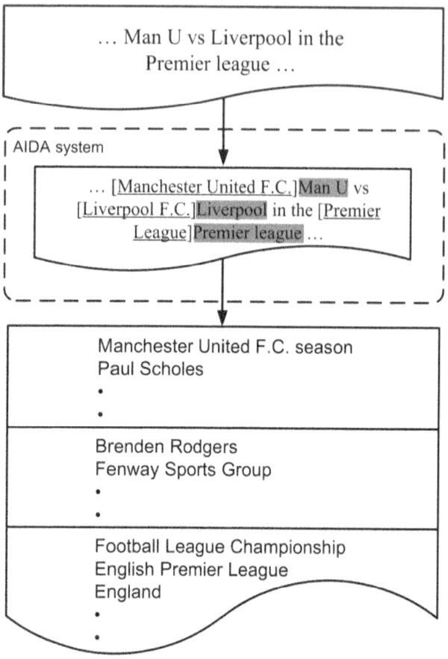

Fig. 3: Sample of generating keyphrases by the system.

```
<entity wiki_title="Parker,_Florida"
type="GPE" id="E0000012" name="Parker,
Florida">

<facts class="Infobox Settlement">
<fact name="official_name">
Parker, Florida
</fact>

<fact name="subdivision_name">
<link entity_id="E0679687">United States</
link>
</fact>

<fact name="subdivision_name1">
<link entity_id="E0373950">Florida</link>
</fact>
......
</facts>

<wiki_text>
<![CDATA[Parker, Florida
Parker is a city in Bay County, Florida,
United States [...] 4.6% of those age 65 or
over.
]]>
</wiki_text>

</entity>
```

Fig. 4: Sample KB candidate entity page containing a set of facts and its informative context.

Given that there are several thousands of weighted keyphrases for each entity in YAGO2, a keyphrase weight threshold (set to 0.002) was manually determined for filtering out the less reliable (all keyphrases with weight less than 0.002 in YAGO2) and getting a smaller and more focused set of keyphrases. In general, our system extracts ~300 keyphrases for each entity mention in the background document. Figure 3 shows an example of using AIDA to exploit keyphrases related to following mentions “Man U,” “Liverpool,” and “Premier league” occurring in the background document.

Alternate Name Generation. Generating Alternate Names (AN) of each query can effectively reduce the ambiguities of the mention, under the assumption that two name variants in the same document can refer to the same entity. We follow the techniques below to generate AN:

- *Acronym Expansion.* Acronyms form a major part of ORG queries and can be highly ambiguous, e.g. “ABC” is referred to around 100 entities. The purpose of acronym expansion is to reduce its ambiguity. The system seeks inside the background document to gather all subsequent tokens with the first capital orderly matched to the letters of the acronym. Also, the expansions are acquired before or inside the parentheses, e.g. "Congolese National Police (PNC)", or "PNC (Congolese National Police)".

- *Gazetteer-based AN Generation.* Sometimes, query names are abbreviations. In these occasions, auxiliary gazetteers are beneficial to map the pairs of ⟨*abbreviation, expansion*⟩ such as the US states, (e.g. the pair ⟨*CA, California*⟩ or ⟨*MD, Maryland*⟩), and country abbreviations, (e.g. the pairs ⟨*UK, United Kingdom*⟩, ⟨*US, United States*⟩ or ⟨*UAE, United Arab Emirates*⟩).
- *Google API.* In more challenging cases, some query names contains grammatical irregularities or a partial form of the entity name. Using Google API, more complete forms of the query name are probed across the Web. For doing so, the component captures title of first (top ranked) result of the Google search engine as possibly better form of the query name. For instance, in the case of query name "Man U", using the method above, the complete form "Manchester United F.C." is obtained.

3.3 Candidate Generation

Given a particular query, q, a set of candidates, C, is found by retrieving those entries from the KB whose names are similar enough (Figure 2b), using Dice measure, to one of the alternate names of q found by the query expansion. In our experiments we used a similarity threshold of 0.9, 0.8 and 1 for PER, ORG and GPE respectively. By comparing the candidate entity type extracted from the corresponding KB page and query type obtained by NERC, we filter out those candidates having different types to attain more discriminative candidates.

In general, each KB entity page contains three main parts, a) information of the entity (title, type, id, and name), b) facts about the entity, and c) an informative context about the entity (Figure 4). As shown in the figure, these facts might in turn include the id of other relevant entities. The system enriches the context part of each KB candidate by extracting the fact ids to get their corresponding KB pages in order to merge their relevant informative contexts with the current one. By applying this technique, the context of each candidate could be more informative. The mentioned figure shows the KB page corresponding to "Parker, Florida." The module collects the `wiki_text` information of its related entities "United States" and "Florida" to enrich the `wiki_text` of "Parker, Florida."

3.4 Candidate Ranking and NIL Clustering

In EL, query expansion techniques are alike across systems, and KB node candidate generation methods normally achieve more than 95% recall. Therefore, the most crucial step is ranking the KB candidates and selecting the best node.

- *Topic Modeling.* This module sorts the retrieved candidates according to the likelihood of being the correct referent. We employs VSM [21], in which a vectorial representation of the processed background document is compared with the vectorial representation of the candidates' `wiki_text`. The vector space domain consists of the whole set of words within the keyphrases found in the enriched background document and the rank consists of their tf/idf computed against the set of candidates' `wiki_text`. All common words (using stop-list) and all words that appear only once are removed. We use cosine similarity. In addition, in order to reduce dimensionality we apply *Latent Semantic Indexing* (LSI) (Figure 2c). The system selects the candidate having the highest degree of similarity as a correct reference of the query name.
- *Term Clustering.* For those queries referring to entities which are not in the KB (NIL queries), the system should cluster them in groups, each referring to a same Not-In-KB entity. To this objective, a term clustering method is applied to cluster such queries. Each initial NIL query forms a cluster assigning a NIL id. The module compares each new NIL query with each existing cluster (initial NIL query) using a dice coefficient similarity between all ANs (including query name) of both queries. If the similarity is higher than the predefined NIL threshold, the new NIL query is associated to this cluster, otherwise it forms a new NIL cluster. In our experiments we used 0.8 as NIL threshold.

4 Evaluation Framework

We have participated in the framework of the TAC-KBP 2012 and TAC-KBP 2013 mono-lingual EL evaluation tracks[3].

Given a list of queries, each consisting of a name string, a background document, and a pair of character offsets indicating the start and end position of the name string in the document, the system is required to provide the identifier of the KB entry to which the name refers if existing, or a NIL ID if there is no such KB entry. The EL system is required to cluster together queries referring to the same non-

3 http://www.nist.gov/tac/

KB (NIL) entities and provide a unique ID for each cluster. The reference KB used in this track includes hundreds of thousands of entities based on articles from an October 2008 dump of English WP, which includes 818,741 nodes. The evaluation query sets in 2012 and 2013 experiments contain 2229 and 2190 queries respectively. Entities generally occur in multiple queries using different name variants and/or different background documents. Some entities share confusable names, especially challenging in the case of acronyms.

5 Results and Analysis

Previously, we participated in TAC-KBP 2012 and TAC-KBP 2013 EL evaluation tracks. The system presented in this paper is an improved version of the system with which we participated in the TAC-KBP 2013 track. Using the TAC-KBP 2012 and TAC-KBP 2013 evaluation queries, we present our results splitted into four parts: official results of TAC-KBP 2012 and TAC-KBP 2013 named by '2012' [6] and '2013' [1], and results of the improved system named by '2012*' and '2013*'.

Twenty five teams participated and submitted 98 runs to the TAC-KBP English EL evaluation in 2012, and 26 teams submitted a total of 111 runs to the TAC-KBP in 2013. Tables 3a and 3b illustrate the results obtained by our systems (both base-line and improved systems) over the TAC-KBP 2012 and TAC-KBP 2013 EL evaluation frameworks using B-cubed+ metric (including Precision, Recall, and F1). The tables split the results by those query answers existing in reference KB (in-KB) and those not in the KB (NIL). Evaluation corpus in TAC-KBP 2012 includes two kinds of genres, News Wires (NW) and Web Documents (WB). In TAC-KBP 2013, a new genre, Discussion Fora (DF) was associated to the evaluation corpus. DF is highly challenging since it contains many grammatical irregularities extracted from fora, blogs, etc. The tables also indicate the results by three different query types, PER, ORG, and GPE.

In both experiments, the proposed system achieved a significant improvement compared with our base-line results shown in the mentioned tables. In 2012 experiment, as shown in Table 3a, 2012* achieved improvement in all portions over the median of the results achieved by all the participants, except in the case of GPE query types that we attained a little decrement less than median. In addition, in 2013 experiment, as shown in Table 3b, 2013* also improved the performance compared with the previous results of our participation in TAC-KBP 2013. The system obtained a notable result in the case of GPE and grabbed scores a little less than median for PER and ORG. The reason that the results for the same query type varies in 2012 and 2013 is because the nature of queries are different in

Tab. 3: The results comparison between the systems over TAC-KBP2012 (Table 3a) and TAC-KBP2013 (Table 3b) mono-lingual EL evaluation framework.

(a)

System	B^3 + F1							
	All (2226)	in-KB (1177)	NIL (1049)	NW (1471)	WB (755)	PER (918)	ORG (706)	GPE (602)
2012	0.421	0.311	0.545	0.460	0.344	0.599	0.382	0.194
2012*	**0.611**	**0.524**	**0.710**	**0.665**	**0.507**	**0.771**	**0.560**	**0.426**
Median	0.536	0.496	0.594	0.574	0.492	0.646	0.486	0.447
Highest	0.730	0.687	0.847	0.782	0.646	0.840	0.717	0.694

(b)

System	B^3 + F1								
	All (2190)	in-KB (1090)	NIL (1100)	NW (1134)	WB (343)	DF (713)	PER (686)	ORG (701)	GPE (803)
2013	0.435	0.285	0.584	0.508	0.485	0.284	0.535	0.538	0.248
2013*	**0.602**	**0.591**	**0.599**	**0.663**	**0.532**	**0.535**	**0.586**	**0.575**	**0.636**
Median	0.584	0.558	0.603	0.655	0.546	0.458	0.620	0.599	0.526
Highest	0.721	0.724	0.720	0.801	0.673	0.633	0.758	0.737	0.720

these years, e.g. most GPE queries in 2012 are focused on the U.S. states but most GPE queries in 2013 are about countries. The EL task for the evaluation queries in 2013 was increasingly more strict than those presented in 2012, including more ambiguous and partial query names along with a lot of grammatical irregularities. In both experiments, the scores obtained for the NW genre is higher than WB since the NW documents contain more structured text. In addition, in 2013, the scores obtained for DF genre are the lowest compared against NW and WB genres. The reason is that, DF genre includes many typos and grammatical errors. The

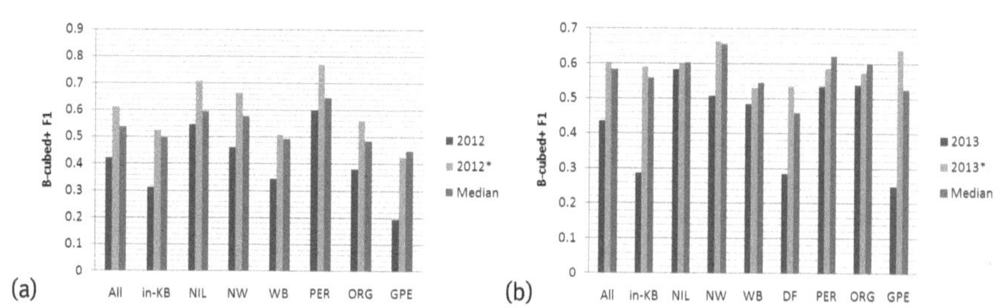

Fig. 5: B^3 + F1† comparison between our systems and the median of the results achieved by all the participants in the 2012 (Figure 5a) and 2013 (Figure 5b) experiments.
†: n.b. the B^3 + metric measures Precision, Recall, and F1 of systems focusing on ability of them to cluster queries.

difference between our overall result and the median in 2012 experiment is higher than its difference in 2013. Since the teams participated in TAC-KBP 2012 and TAC-KBP 2013 are different, thus comparing between the medians in 2012 and 2013 is not possible. In total, thanks to our proposed system we gained the overall scores more than median in both 2012 and 2013 experiments (Figures 5a and 5b).

6 Conclusions and Future Work

The improved version of the system presented here ran over the TAC-KBP2012 and TAC-KBP2013 EL evaluation framework. The results were compared with those obtained by all participants. Most participants have achieved much research in the field of EL having many contributions in this regard. Thus, the results are considered a good framework for comparing the performance of the systems. We achieved the comparison using B-cubed+ F1 metric. The measure depicted a significant improvement compared with our previous results and higher than the median of participant results.

As a future work, we carry on to improve the system results through profound analyzing of semantics of the keyphrases in order to increase the accuracy of the task. The deep analysis of the keyphrases is beneficial when the queries are highly ambiguous to realize their correct references.

Acknowledgments

This work has been produced with the support of the SKATER project (TIN2012-38584-C06-01).

Bibliography

[1] Ageno, A., Comas, P.R., Naderi, A., Rodríguez, H., Turmo, J.: The talp participation at tac-kbp 2013. In: the Sixth Text Analysis Conference (TAC 2013), Gaithersburg, MD USA (2014)
[2] Banerjee, S., Pedersen, T.: Extended gloss overlaps as a measure of semantic relatedness. IJCAI **3** (2003)
[3] Bunke, H., Alberto Sanfeliu, e.: Syntactic and structural pattern recognition: theory and applications. World Scientific **7** (1990)
[4] Fellbaum, C.: Wordnet: An electronic lexical database. MIT Press (1998)
[5] Fu, K.S.: Syntactic pattern recognition and applications. Prentice-Hall (1982)

[6] Gonzàlez, E., Rodríguez, H., Turmo, J., Comas, P.R., Naderi, A., Ageno, A., Sapena, E., Vila, M., Martí, M.A.: The talp participation at tac-kbp 2012. In: the Fifth Text Analysis Conference (TAC 2012), Gaithersburg, MD USA (2013)

[7] Hoffart, J., Suchanek, F.M., Berberich, K., Weikum, G.: Yago2: a spatially and temporally enhanced knowledge base from wikipedia. Artificial Intelligence Journal (2013)

[8] Hoffart, J., Yosef, M.A., Bordino, I., Furstenau, H., Pinkal, M., Spaniol, M., Taneva, B., Thater, S., Weikum, G.: Robust disambiguation of named entities in text. In: the EMNLP Conference, Scotland. (2011)

[9] John, G.H., Langley, P.: Estimating continuous distributions in bayesian classifiers. In: the Eleventh conference on Uncertainty in artificial intelligence, Morgan Kaufmann Publishers Inc. (1995)

[10] Klivans, A.R., Servedio., R.A.: Toward attribute efficient learning of decision lists and parities. The Journal of Machine Learning Research **7** (2006) 587–602

[11] Lesk, M.: Automatic sense disambiguation using machine readable dictionaries: how to tell a pine cone from an ice cream cone. In: the 5th annual international conference on Systems documentation, ACM (1986)

[12] Lin, D.: Automatic retrieval and clustering of similar words. In: the 17th international conference on Computational Linguistics. Volume 2., Association for Computational Linguistics (1998)

[13] Magee, J.F.: Decision trees for decision making. Graduate School of Business Administration, Harvard University (1964)

[14] McCulloch, W.S., Pitts, W.: A logical calculus of the ideas immanent in nervous activity. The bulletin of mathematical biophysics **5**(4) (1943) 115–133

[15] Mihalcea, R.: Co-training and self-training for word sense disambiguation. In: the Conference on Natural Language Learning. (2004)

[16] Mooney, R.J.: Comparative experiments on disambiguating word senses: An illustration of the role of bias in machine learning. In: Conference on Empirical Methods in Natural Language Processing (EMNLP). (1996) 82–91

[17] Navigli, R.: Word sense disambiguation: A survey. ACM Computing Surveys (CSUR) **41**(2) (1990)

[18] Pantel, P., Lin, D.: Discovering word senses from text. In: the 8th ACM SIGKDD international conference on Knowledge discovery and data mining, ACM (2002)

[19] Quinlan, J.R.: C4. 5: programs for machine learning. Volume 1. Morgan Kaufmann (1993)

[20] Ratinov, L., Roth, D.: Design challenges and misconceptions in named entity recognition. In: CoNLL. (2009)

[21] Řehůřek, R., Sojka, P.: Software Framework for Topic Modelling with Large Corpora. In: Proceedings of the LREC 2010 Workshop on New Challenges for NLP Frameworks, Valletta, Malta, ELRA (May 2010) 45–50 http://is.muni.cz/publication/884893/en.

[22] Rivest, R.L.: Learning decision lists. Machine learning **2**(3) (1987) 229–246

[23] Schutze, H.: Dimensions of meaning. In: Supercomputing '92: ACM/IEEE Conference on Supercomputing, IEEE Computer Society Press (1992) 787–796

[24] Towell, G., Voorhees, E.M.: Disambiguating highly ambiguous words. Computational Linguistics **24**(1) (1998) 125–145

[25] Widdows, D., Dorow, B.: A graph model for unsupervised lexical acquisition. In: the 19th international conference on Computational linguistics. Volume 1., Association for Computational Linguistics (2002)

Paweł Chrząszcz

Extraction of Polish Multiword Expressions

Abstract: Natural language processing for fusional languages often requires using inflection dictionaries which lack multiword expressions (MWE) – proper names and other sequences of words that have new properties as a whole. This paper describes an effort to extract such MWEs from Polish text. The search is not limited to any particular domain and there are no predefined categories, gazetteers or manually defined rules, what makes it different from named entity recognition (NER). As there are no Polish linguistic resources containing MWEs, we cannot use supervised learning techniques, so Wikipedia content and link structure are used to create syntactic patterns that are recognised in text using non-deterministic automata. The results can be used directly or transformed into an inflection dictionary of MWEs. Several algorithms are presented as well as results of a test on a corpus of Polish press releases.

1 Introduction

Natural language processing usually makes use of either simple word frequency statistics or more advanced language-dependent features. For Polish (and other inflecting languages) such features are: part of speech (POS), lemma and grammatical tags: gender, case etc. Feature extraction can be done using statistical algorithms (Kuta et al. 2007) or with a morphological analyser such as Morfeusz (Woliński 2006) or Morfologik[1]. Alternatively, one can use the Polish Inflection Dictionary (SFJP)[2]. The dictionary is still being improved and current effort aims at extending it with semantic information – full semantic relations are introduced

1 http://morfologik.blogspot.com
2 SFJP – a Polish dictionary developed by the Computer Linguistics Group at AGH University of Science and Technology in Kraków, in cooperation with the Department of Computational Linguistics at the Jagiellonian University (Lubaszewski et al. 2001). It contains more than 120 thousand headwords and provides a programming interface – the CLP library (Gajęcki 2009).

Paweł Chrząszcz: Computational Linguistics Department, Jagiellonian University, ul. Gołębia 24, 31-007 Kraków, and Computer Science Department, AGH University of Science and Technology, Aleja Adama Mickiewicza 30, 30-059 Kraków, e-mail: p.chrzaszcz@uj.edu.pl

Tab. 1: Examples of nominal MWEs that we focus on.

Word type	Examples
Person names	Maciej Przypkowski, Allen Vigneron, Szymon z Wilkowa (*Szymon from Wilków*)
Other proper names	Lazurowa Grota (*Azure Cave*), Polski Związek Wędkarski (*Polish Fishing Association*)
Other named entities	rzeka Carron (*River Carron*), jezioro Michigan (*Lake Michigan*), premier Polski (*Prime Minister of Poland*)
Terms of art	martwica kości (*bone necrosis*), dioda termiczna (*thermal diode*), zaimek względny (*relative pronoun*)
Idioms and other common words	panna młoda (*bride*), piłkarz ręczny (*handball player*), baza wojskowa (*military base*)

for as many words as possible (Pohl 2009, Chap. 12.2) while the rest of them should have short semantic labels (Chrząszcz 2012). However, SFJP, as well as other inflection resources for Polish, is missing **multiword expressions** (MWEs). They are lexemes that consist of multiple tokens and have properties that differ from what can be inferred from individual tokens. We could roughly define them as "idiosyncratic interpretations that cross word boundaries" (Sag et al. 2002). MWE often has a completely new meaning: "panna młoda" means *bride*, while literal meaning would be *young maid* (Lubaszewski 2009). Proper names are also often MWEs. Their meaning and inflection should definitely be present in inflection dictionaries to allow their recognition and processing.

The goal of this paper is to present several methods of automatic extraction of MWEs from Polish text. The algorithms use the Polish Wikipedia to learn **syntactic patterns** that are later recognized in raw text. The results include inflection information and can either be used directly or transformed into an inflection dictionary of MWEs. We focus on nominal MWEs (table 1), because the majority of MWEs are nouns and Wikipedia headwords are nominal phrases.

1.1 Related Work

A widespread type of multiword recognition is **Named Entity** (NE) **Recognition** (NER). NEs are phrases that represent entities belonging to predefined categories: people, geographical objects, dates etc. NER often makes use statistical methods e.g. maximum entropy model (Tjong Kim Sang and De Meulder 2003) or HMMs that can be combined with predefined syntactic and semantic rules as well as gazetteers, e.g. Zhou and Su (2002) for English, Piskorski et al. (2004) for Polish. Wikipedia has also been used as a source for NER. One could use links and definitions (Nothman et al. 2009) or categories and interwiki links (Richman and Schone 2008) as the training data. Although these attempts often reached F1 (F-measure) of 70–90% on various corpora, they were limited to carefully selected NE categories (usually the ones giving best results) and used manual rules or gazetteers. As our goal is to detect all nominal MWEs without any rules or limits, we must look for a more general approach.

For some languages, e.g. French, MWEs are included in dictionaries and corpora (Constant et al. 2012), what makes it possible to recognise them with F1 over 70%. For other languages, one could use association measures such as Mutual Information (MI), χ^2 or Permutation Entropy (Zhang et al. 2006). Ramisch et al. (2008) evaluated these methods on simple structures (e.g. adjective-noun), but although they used language-dependent information, precision remained low. Attia et al. (2010) used Wikipedia to extract nominal semantic non-decomposable MWEs relying on Crossligual Correspondence Asymmetries. The extraction results were promising with F1 over 60%, but there was no recognition test on a corpus. Finally, Farahmand et al. (2014) used a manually tagged sample of 10K negative and positive English MWE examples to train an SVM classifier. The results were very positive (89% F1) but the test was performed only on the predefined examples. Our approach differs from the previous attempts as no MWE resources are used, inflection information is extracted, we do not limit the nominal MWEs in any way and we test various algorithms by tagging a corpus sample.

1.2 Anatomy of Polish Nominal MWEs

A multiword expression consists of two or more *tokens*: words, numbers or punctuation marks. A token that is a Polish word can be either fixed or inflectable. Inflectable words form the core of the MWE. They are mostly nouns and adjectives, but can also be numerals and participles. In the base form, all inflectable tokens must occur in nominative case. The number is usually singular, but there are also some *pluralia tantum* – all inflectable tokens must be plural for them. The inflectable tokens need not have the same gender, e.g. "kobieta kot" (*catwoman*).

Example: "Związek Lekkoatletyczny Krajów Bałkańskich" (*Athletic Association of Balkan Countries*). This expression begins with an inflectable non-animate male noun "związek" followed by an inflectable non-animate male nominative adjective "lekkoatletyczny", followed by two fixed words: a noun and an adjective, both in genitive case. The instrumental case of this expression would be "Związkiem

Lekkoatletycznym Krajów Bałkańskich" – inflectable tokens change their form accordingly. We consider it illegal to change the gender of any word or to modify their order. We use SFJP for word recognition.

2 MWE Extraction Methods

2.1 Extraction Using the Wikipedia as a Dictionary

The simplest way to use the Wikipedia for MWE recognition is to treat it as a dictionary. We filtered out headwords with only one token as well as the ones for which we could not extract semantic labels (Chrząszcz 2012). The remaining headwords were used to create patterns that consist of one entry for each token. For words that can be inflectable, the pattern contains the word ID (from SFJP) and the grammatical categories[3]. For fixed tokens (punctuation, words not in SFJP, prepositions, verbs etc. as well as words in non-nominative cases) the pattern contains just the token itself.

Example: For the headword "Związek Lekkoatletyczny Krajów Bałkańskich" the pattern is shown below in a simplified form. '?' means *maybe inflectable*. It is not possible to tell if "związek" should be capitalised because all Wikipedia headwords are capitalised.

> [**? (z|Z)wiązek** *male non-animate singular noun*] [**? Lekkoatletyczny** *male singular adj.*] [**Krajów**] [**Bałkańskich**]

2.1.1 Dictionary Pattern Matching (DM)

To check if a sequence of tokens matches a pattern, we created a non-deterministic finite automaton with one transition per pattern option for each token. If a pattern entry for a token is ambiguous, the automaton forks into multiple states and its diagram becomes a tree.

Each matching step means checking all the possible transitions for the current state set and the current input token. For a fixed token, it is enough to check if the token is the same as the pattern entry. For an inflectable word though, the matching process is more complex. One has to check if the word ID is the

3 Actually there are can be multiple values for both the word ID and the categories because of language ambiguity.

same and if the form is a valid inflected one. This would e.g. eliminate the phrase "Związkiem Lekoatletyczną Krajów Bałkańskich" (second token changed gender to female). The form has to agree for all inflected words, so for the first of them the grammatical categories (number, case) are extracted and they need to be the same for all subsequent inflected tokens. This would eliminate e.g. "Związkiem Lekkoatletycznemu Krajów Bałkańskich" (first token in instrumental case, second in accusative case).

The result of traversing the above automaton is a list of visited accepting states (alternatively: a list of paths leading to these states). This concept can be expanded to allow matching multiple patterns at once – all automata can be combined into one. Its output will be a list of *(pattern, path)* pairs. This automaton can be then applied to raw text by starting it for each subsequent word. Whole corpus can be then analysed in one pass by performing all the matching simultaneously.

The desired output format is a list of matching words with their declension information. We decided not to allow overlapping words. The tests revealed that in these cases choosing a non-overlapping subset covering as many tokens as possible is the best option.

2.2 Extraction Using Syntactic Patterns

The performance of DM is limited to the headwords of Wikipedia, so that method is incapable of recognising any unknown words. One can observe that there are groups of MWEs which share similar syntactic structure, e.g. "tlenek węgla" (*carbon dioxide*) and "chlorek sodu" (*sodium chloride*) both contain one inflectable male noun and one fixed male noun in genitive case. What is more, they both often appear in the text in similar context – e.g. in genitive case after a noun ("roztwór chlorku sodu" – *a solution of sodium chloride*, "reakcja tlenku węgla" – *a reaction of carbon dioxide*) or after the preposition "z" (*with*) in instrumental case ("reakcja … z chlorkiem sodu" – *a reaction of … with sodium chloride*). We decided to automatically create **syntactic patterns** that would express these regularities and recognise them in text using an automaton similar to the one used for DM. The matching algorithm is called **Syntactic pattern Matching (SM)**. The patterns should not be created by hand – otherwise they would be limited to the structures known to their author. We tested several possible pattern structures and chose the one presented below. For each token of the MWE the pattern contains one entry:

- For inflectable words, it begins with an asterisk (*) and contains the POS and (when possible) gender, number and case of the base form. Each of these features can be ambiguous.

- For prepositions and some punctuation marks, the token itself is the pattern.
- For fixed words that are in SFJP, it contains the POS and (when possible) gender and case. Each of these features can be ambiguous.
- For other tokens it contains the category of the token (word, number, punctuation, other).

Example. The pattern for "czarna dziura" (*black hole*) is:

[* *fem. singular noun*] [* *fem. singular adj.*]

To include the information about the context, we decided that each pattern may be connected to one or more **context patterns** that have a similar structure to the core pattern (of course context tokens are not inflectable). The context covers one word before and one after the MWE. For example, if "czarna dziura" occurs in the expression: "(...) masywnej **czarnej dziury**.", the context pattern will be:

[*fem. singular noun, gen., acc. or loc.*] **[* *fem. singular noun*] [* *fem. singular adj.*]**[punct.]

2.2.1 Pattern Construction

For each headword from Wikipedia we need to know which tokens are inflectable, how they are inflected (for ambiguous words), in what context they appear in text and if the first word is lowercase. Our choice was to gather this information from links from other articles leading to the given headword as well as occurrences of the inflected word in the article itself. Having a list of inflected occurrences of the MWE with their contexts, inferring the inflection pattern does not seem complicated, but the algorithm is actually fairly complex. This is due to the ambiguities of fusional languages and errors present all over Wikipedia. The algorithm thus has to find the largest subset of occurrences that lead to a non-contradictory pattern and has exponential complexity, so we implemented several optimisations that successfully limit the impact of this fact.

Because we know how many times each context pattern occurred in each grammatical form, the automaton performing SM can calculate a **score** of each matched phrase: it would be the total number of all occurrences of the matched patterns for the recognised form[4]. We use the scores to choose the best among overlapping expressions and as a threshold used to filter out low quality matches.

4 Actually we operate on form sets and their intersections because the language is ambiguous.

2.3 Compound Methods

In an attempt to improve recognition performance, one can combine the two methods described above. The most important possibilities are listed below. Some of them are using Wikipedia Corpus (WC) that we created from the whole Wikipedia text. We also tried an iterative approach, but it did not provide significant improvement.

DM for words with patterns (pDM) Like DM, but the input headwords are limited to the ones for which syntactic patterns were created. This allows identifying inflectable tokens and lowercase first tokens. The matches are scored according to the number of occurrences of the pattern for the given word. The main drawback is that the number of words is reduced.

DM for SM results (SM-DM) Like pDM, but uses a second dictionary, created from the results of SM on WC. To obtain this dictionary, SM results are grouped by lemma (they also need some disambiguation and conflict resolution). A concept of **word score** is introduced – it is the sum of all scores of the contributing matches. SM-DM provides scores for results – they are pDM scores from the first dictionary and word scores from the second one. This method significantly increases recall at the cost of precision.

SM for SM results (SM-SM) Like SM, but uses a second set of patterns. It is created from the results of SM on WC. A dictionary is constructed (like for SM-DM) and the syntactic patterns are created from that dictionary. The goal of the first SM phase is to find more words that can be used to create patterns.

SM for pDM results (pDM-SM) Like SM-SM, but the dictionary is created using pDM, not SM. The pDM phase aims at providing additional context patterns for SM.

SM for DM results (DM-SM) This method is an attempt to create the syntactic patterns using the whole Wikipedia – a dictionary is created from the results of DM on WC and it is used to create the patterns for SM. The main drawback is that a lot of inflection ambiguities remain from the DM phase.

3 Tests and Results

There are no known corpora of Polish text tagged for MWEs, so we needed to create the test set from a corpus. We already had the Wikipedia Corpus, but cutting a part of Wikipedia was problematic as its network structure (which was used by the methods in multiple places) was broken. As a result, we decided to use a corpus of press releases of the Polish Press Agency (PAP), as it is a large collection of texts from multiple domains. 100 randomly chosen press releases were tagged manually by two annotators. All disagreements were discussed and resolved. An example of a tagged release is given below:

> *We wtorek* ***[*mistrz *olimpijski]****, świata i Europy w chodzie* ***[*Robert *Korzeniowski]*** *weźmie udział w stołecznym* ***[*Lesie *Kabackim]*** *w* ***[*Biegu ZPC SAN]*** *o* ***[*Grand *Prix]*** *Warszawy oraz o* ***[*Grand *Prix] [*Polskiego *Związku Lekkiej Atletyki]****.*

The were two types of the test: the **structural test** required only the tokens of the MWE to be the same for the expected and matched expressions and the **syntactic test** also required the inflectable words to be correctly indicated. All parameters of the tested methods were optimised using a separate test set of 10 press releases – the main test sample was used only for the final evaluation.

3.1 Test Results

Table 2 shows results optimized for best F1 score of the structural test. Basic dictionary algorithms have high precision but low recall, because Wikipedia does not contain all possible MWEs. DM has poor syntactic test results because the inflection patterns are unknown. SM0 is SM without the pattern contexts. SM preforms a bit better because of the contexts, but it is limited by sparse link connections in Wikipedia. The only compound method that brings significant improvement is SM-DM. Fig. 1 (left) presents precision-recall graphs for the best performing methods.

All the errors for the best performing method, **SM-DM**, were analysed one by one. It turned out that in many cases the identified phrases are not completely wrong. Such problematic cases are:

Tab. 2: Results of the test on a sample of 100 tagged PAP press releases.

	Structural test			Syntactic test		
Algorithm	**Precision**	**Recall**	**F1**	**Precision**	**Recall**	**F1**
SM-DM	65.87	65.20	65.53	57.12	56.54	56.83
SM-SM	52.99	55.69	54.30	44.59	46.86	45.70
pDM-SM	50.07	57.39	53.48	42.22	48.39	45.09
SM	49.93	57.39	53.40	42.10	48.39	45.02
SM0	49.85	56.54	52.98	41.32	46.86	43.91
DM	75.51	37.69	50.28	44.22	22.07	29.45
pDM	90.72	29.88	44.96	86.08	28.35	42.66
DM-SM	47.59	25.13	32.89	38.59	20.37	26.67

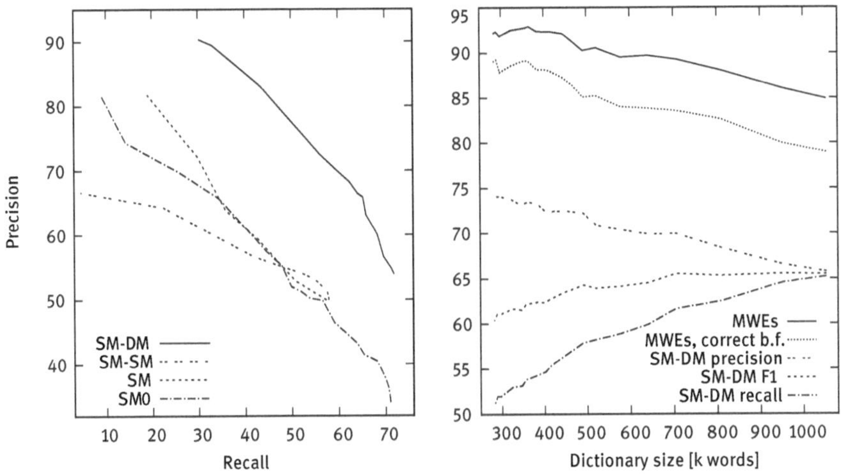

Fig. 1: Left: precision-recall graphs for the structural test. **Right**: Quality of the dictionary created by the SM phase of SM-DM as a function of changing dictionary size after applying different word score thresholds. From top to bottom: % of true MWEs in the dictionary created by the SM phase, % of true MWEs with correctly identified base forms, precision, F1 and recall of SM-DM.

- Overlapping expressions difficult to choose from (even for a human): “Piłkarska reprezentacja Brazylii” contains two MWEs: a term “reprezentacja piłkarska” with altered token order (*a football team*) and a named entity “reprezentacja Brazylii” (*Brazilian team*).
- Unusually long names, spelling and structural errors: “Doroty Kędzierzwskiej” (misspelled surname), “Polska Fundacja Pomocy humanitarnej ‘Res Humanae’” (unusual structure and an accidental lowercase letter).
- Phrases that are not MWEs in the particular (semantic) context: “osoba paląca sporadycznie” contains the term “osoba paląca” (*a smoker*) while it means *a person smoking sporadically*.
- For the inflection errors: one of the words is not in SFJP (usually a surname): “Janusz Steinhoff”, “Władysław Bartoszewski”.

The SM phase of SM-DM produces a dictionary of 4.45 million words that can be filtered by score to achieve either high precision or recall. The threshold of 150 yielded best results (65.5% F1 for SM-DM), so we analysed the quality of the dictionary for a random sample of 1000 entries above that score. About **79%** of the resulting **1.05** million words were correct MWEs and further **6%** were MWEs, but had wrong base forms. This quality can be further increased at the cost of dictionary size and SM-DM recall (fig. 1, right). Although it is difficult to estimate the target size of an MWE lexicon as it is unclear how large human lexicons are (Church 2013), the results make us consider SM a good method for building MWE lexicons.

3.2 Future Work

There seems to be some significant room for further improvement, especially in the SM algorithm. The first step would be to eliminate the errors that are results of words missing from SFJP either by adding another dictionary (e.g. Morfologik) or by filling missing people (sur)names directly in SFJP. We also remind that one of our assumptions was not to use any user-defined rules. However, there were many false positives in the dictionary that could be eliminated by manual patterns, e.g. "wrzesień 2011" (*September 2011*), "część Sztokholmu" (*a part of Stockholm*), "pobliże Jerozolimy" (*vicinity of Jerusalem*). This could improve the dictionary quality and increase the precision of SM-DM accordingly. We could also use other Polish corpora instead of the Wikipedia content. The tests could also be improved to allow overlapping words (they are not necessarily errors). After improving the algorithms, we plan to incorporate the resulting MWEs into SFJP and transfom SM-DM into an integral part of this dictionary.

4 Conclusions

The presented algorithms are the first successful attempt (known to the authors) of automatic extraction of all Polish MWEs. It is not easy to compare the results to other papers on NER or MWE recognition as they either limit the scope of MWEs to specific categories or use resources that do not exist for Polish. We would like to continue our research and improve the quality of the algorithms as well as the tests. Finally, we would like to extend SFJP with MWEs. Our opinion is that the research is going in the right direction and the final goal is likely to be reached in the near future.

Bibliography

Attia, M., Tounsi, L., Pecina, P., van Genabith, J., and Toral, A. (2010) Automatic extraction of arabic multiword expressions. In: Proceedings of the Workshop on Multiword Expressions: from Theory to Applications, Beijing, China, 28 August 2010, pp. 18–26.

Chrząszcz, P. (2012) Enrichment of inflection dictionaries: automatic extraction of semantic labels from encyclopedic definitions. In: Sharp, B. and Zock, M. (eds.), Proceedings of the 9th International Workshop on Natural Language processing and Cognitive Science, Wrocław, Poland, 28 June – 1st July 2012, pp. 106–119.

Church, K. (2013) How many multiword expressions do people know? ACM Transactions on Speech and Language Processing (TSLP), 10(2):4.

Constant, M., Sigogne, A., and Watrin, P. (2012) Discriminative strategies to integrate multiword expression recognition and parsing. In: Proceedings of the 50th Annual Meeting of the Association for Computational Linguistics: Long Papers – Volume 1, Jeju, Korea, 8–14 July 2012, pp. 204–212.

Farahmand, M. and Martins, R. (2014) In: A supervised model for extraction of multiword expressions based on statistical context features. Proceedings of the 10th Workshop on Multiword Expressions, Gothenburg, Sweden, 26–27 April 2014, pp. 10–16.

Gajęcki, M. (2009) Słownik fleksyjny jako biblioteka języka c. In: Lubaszewski, W. (ed.), Słowniki komputerowe i automatyczna ekstrakcja informacji z tekstu, pp. 107–134. Wydawnictwa AGH, Kraków.

Kuta, M., Chrząszcz, P., and Kitowski, J. (2007) A case study of algorithms for morphosyntactic tagging of polish language. Computing and Informatics, 26(6):627–647.

Lubaszewski, W. (2009) Wyraz. In: Lubaszewski, W. (ed.), Słowniki komputerowe i automatyczna ekstrakcja informacji z tekstu, pp. 15–36. Wydawnictwa AGH, Kraków.

Lubaszewski, W., Wróbel, H., Gajęcki, M., Moskal, B., Orzechowska, A., Pietras, P., Pisarek, P., and Rokicka, T. (2001) (ed.), Słownik Fleksyjny Języka Polskiego. Lexis Nexis, Kraków.

Nothman, J., Murphy, T., and Curran, J. R. (2009) Analysing wikipedia and gold-standard corpora for ner training. In: Proceedings of the 12th Conference of the European Chapter of the Association for Computational Linguistics, Athens, Greece, 30 March – 3 April 2009, pp. 612–620.

Piskorski, J., Homola, P., Marciniak, M., Mykowiecka, A., Przepiórkowski, A., and Woliński, M. (2004) Information extraction for polish using the sprout platform. In: Kłopotek, A., Wierzchoń, T. and Trojanowski, K. (eds.), Proceedings of the International IIS: IIPWM' 05 Conference, Gdańsk, Poland, 13–16 June 2005, pp. 227–236.

Pohl, A. (2009) Rozstrzyganie wieloznacznołści, maszynowa reprezentacja znaczenia wyrazu i ekstrakcja znaczeń In: Lubaszewski, W. (ed.), Słowniki komputerowe i automatyczna ekstrakcja informacji z tekstu, pp. 241–255. Wydawnictwa AGH, Kraków.

Ramisch, C., Schreiner, P., Idiart, M., and Villavicencio, A. (2008) An evaluation of methods for the extraction of multiword expressions. In: Proceedings of the LREC Workshop – Towards a Shared Task for Multiword Expressions, Marrakech, Morocco, 1 June 2008, pp. 50–53.

Richman, A. E. and Schone, P. (2008) Mining wiki resources for multilingual named entity recognition. In: Proceedings of the 46th Annual Meeting of the Association for Computational Linguistics: Human Language Technologies, Columbus, OH, USA, 15–20 June 2008, pp. 1–9.

Sag, I. A., Baldwin, T., Bond, F., Copestake, A., and Flickinger, D. (2002) Multiword expressions: A pain in the neck for nlp. In: Gelbukh, A. (ed.), Proceedings of the Third International Conference on Computational Linguistics and Intelligent Text Processing, Mexico City, Mexico, 17–23 February 2002, pp. 1–15.

Tjong Kim Sang, E. F. and De Meulder, F. (2003). Introduction to the conll-2003 shared task: Language-independent named entity recognition. In: Daelemans, W. and Osborne, M.

(eds.), Proceedings of the Seventh Conference on Natural Language Learning at HLT-NAACL 2003, 27 May – 1 June 2003, Edmonton, Canada, Vol. 4, pp. 142–147.
Woliński, M. (2006) Morfeusz – a practical tool for the morphological analysis of polish. Advances in Soft Computing, 26(6):503–512.
Zhang, Y., Kordoni, V., Villavicencio, A., and Idiart, M. (2006) Automated multiword expression prediction for grammar engineering. In: Proceedings of the Workshop on Multiword Expressions: Identifying and Exploiting Underlying Properties, July 2006, Sydney, Australia, pp. 36–44.
Zhou, G. and Su, J. (2002) Named entity recognition using an hmm-based chunk tagger. In: Proceedings of the 40th Annual Meeting on Association for Computational Linguistics, 7–12 July 2002, Philadelphia, PA, USA, pp. 473–480.

Nabil Abdullah and Richard Frost

Beyond Classical Set

Abstract: This work presents a novel framework for natural language semantics by augmenting the notion of the conventional set using types. Using this newly conceived set, termed "typed set", we demonstrate the viability of this notion by showing how the meaning of non-extensional constructs can be represented and computed without resorting to metaphysically murky, psychologically unappealing and computationally expensive notions.

1 Introduction

Set theory has established its place as *de facto lingua franca* not only in mathematics but in many other realms, including the analysis of natural-language (NL). In NL semantics the process is not straightforward. Because of sets' extensional character and because intension in NL expressions is the rule rather than the exception, many NL scholars drop the notion of sets. Others opt to circumvent the limitation of set theory by appealing to other notions. Carnap (1947) chose to follow such a route. Realizing the shortcoming of sets, he rendered Frege's Sense (*Sinn*) as functions from possible worlds (PWs) to extensions. This move poses no threat to set theory's foundational role, however. Indeed, it can be seen as an evidence of the versatility of sets. For functions – Carnap called them "intensions" – are also set-theoretically represented as special relations on sets. Thus, following Carnap's lead, set-theoretic semantics for non-extensional NL expressions becomes viable using machinery available to extensional systems, without abandoning set theory. The introduction of PWs has led to impressive developments in the semantics of formal languages, e.g., modal logic, and NL culminating in Montague Semantics – "the pinnacle" of formal semantics, as described by Abbott (2010). So, mathematically not much has changed as a result. The price of Carnap's solution is metaphysical: the ontology had to make room for PWs.

Yet, the notion of PWs is not problem-free. Many scholars think that it is problematic – see McGinn (2000:69) for metaphysical concerns, Partee (1979) for psycholinguistic concerns, Cherichia and Turner (1988) for semantic concerns

Nabil Abdullah, Richard Frost: The University of Windsors, 401 Sunset Ave, Windsor, ON N9B 3P4, Canada, e-mail: {nabil, richard}@uwindsor.ca

and, Hirst (1988), Woods (1985) and Rapaport (2013) for computational/cognitive concerns.

Now, we are not so much concerned with PWs but our concern is the root of the problem: the conventional set is an inadequate tool for representing the rich structure of NL expressions. Strengthening it by appealing to other notions, such as PWs, solves some problems but creates new ones. Although set theory stresses the notion of aggregate – a permissive condition to what is to be a whole – it is restrictive on what can be admitted as a member. For only concepts with definite boundaries are suitable for set treatment. This is a clear bias toward mathematical and scientific concepts, while the majority – those which belong to the pre-theoretic sphere – are left out. This outlook seems to be in agreement with the view that only well-defined concepts are worthy of serious study.

However, the shortcomings of the notion of set are not limited to the pre-theoretic discourse. It is well known that sets cannot satisfactorily accommodate the most central notion in philosophy and logic: predication. This is noted by many scholars, e.g., Cocchiarella (1988) and Jespersen (2008). Also, it is inconceivable, according to the Cantorian conception of set, to talk about one object being a member in the same set in two distinct capacities or an object being partially a member of a set. Such talk runs afoul of the view that: a) sets are extensional objects first and foremost; b) set membership is a yes-no affair. These pose no problem for the development of mathematics, as mathematical objects are abstract and are well-defined and if this were not the case mathematicians and logicians would strive to make them so. Consequently, these objects have their properties absolutely and eternally. For instance, it would be anomalous to describe a number as being "former even" or "potentially prime". This leads to the ideal situation where the meaning of a logico-mathematical term is its referent and the meaning of a predicate is the set (or class) of objects which fall under the predicate. Scholars who have qualms about abstracts/universals welcome this use of sets. It is seen to provide a better solution to the problem of properties than the postulation of universals. For in such a view, talk about properties is reduced to talk about sets, which are well understood and have well-defined identity criterion – a fitting solution that is consistent with Quine's maxim of *no entity without identity*. But this extreme view alienates other fields such as AI and Semantics where mental representation, commonsense reasoning and other aspects of meaning are at issue. Consequently, alternatives such as Semantic Networks and Inheritance systems are investigated.

We believe the concept of a set is invaluable despite the above-mentioned shortcomings. We believe the conception can be shored up by generalizing it using types. In doing so we hope to bridge the gap between the well-understood realm of set theory and first-order logic, on one hand, and the sophistication which is required to go beyond mere extensional aspects of both set theory and first-order logic.

2 Typed Sets

The main claim of this paper is this: *urelements* are better conceived as non-simple. This claim may appear reactionary when viewed against the backdrop of the major developments which has taken place in recent centuries. This includes the abandonment of the notion of axiom simpliciter – after the discoveries of non-Euclidian geometries – and the dethroning of Aristotelian Essentialism and the consequent ascendance of Extensionalism in its stead. Consequently, the formalist's attitude is to seek an austere ontology or even better "desert landscapes", as Quine (1948) puts it. So set-theorists assume nothing except the empty set at the foundation of their axioms. Thus, an urelement is simple, unanalyzable and nothing more could be asked for. This outlook ties well with the view that "Mathematics is the science of relations. What matters is the relation between objects, not the objects themselves", Bombieri (2010). But, here lies a hidden assumption: entities which are the subject of mathematical investigation are given, complete and stable. The task of investigating the composition of entities, if any, is left for metaphysics and empirical sciences. A metaphysician may wonder whether a particular is metaphysically simple or is made up of a bundle of yet simpler constituents – and whether these are "tied up" by some bare substratum or not – and how abstract and concrete particular may differ; and the physicist may investigate sub-atomic particles and whether there is an absolutely simple entity at all.

We believe it is instructive to postulate that urelements are not as simple. With this view, an urelement is a composite, consisting of two parts: a "core" and a type. We believe this conception is of great utility: a finer granularity of representation can be achieved. We demonstrate that this can be done without undermining the conventional conception of set. In fact, we claim it is a natural extension of the Cantorian set. It further endows sets with much more expressive power which goes beyond the confinements of classical set. Indeed, it covers conceptual landscape that neither fuzzy sets nor classical sets do. For typed sets are still bivalent, *pace* the former, yet they allow more grey areas, *pace* the latter.

In the following subsections, we introduce typed sets. Therein, we use the term "type" in the abstract, theory-free sense. In §4, where typed sets are used within the semantics of natural language, we spell out how types must be rendered in a more concrete, theory-specific terms. At this point, however, it suffices to say that in a logical system every non-logical expression designates an object where a type is part and parcel of the object.

2.1 Notion and Notation

We conceive of an urelement as consisting of a core and a type. The core can exhibit many types. The type can be thought of as the object's interface. It might be helpful to draw an analogy: an urelement can be thought of as a rotary rolodex. At some particular time a rolodex can be in one of two states: open (and presenting some contact information) or closed. In both cases it is still the same rolodex. Similarly, a member can participate in some type – corresponding to an open rolodex or it can participate in the generic type (see below) – analogous to that of a closed rolodex. In all cases, it is still the very same member.

The urelement, in addition to being part of an aggregate, can also participate in the type of the containing set. Although set members can be of the same type as the set, their types are distinct. What makes them of the same type is their participation in the type, i.e., that of the containing set. Our position has undeniable affinity to that of the Trope theorist, Loux (2006:71), where attributes are considered *particulars* – as opposed to *universals* – and a term, which the realist takes to stand for a universal, names a set of resembling tropes. However, while the trope theorist talks of set – without qualification – we talk of *typed* set.

The notation we adopt is simply that of ZF set. In particular, we adopt the conventional set notation of braces to represent sets. Sets' type are designated by capital Greek, while members' by small Greek letters or a letter juxtaposed by a sequence of primes. This is to encode the member-member and member-set type similarity and distinction just mentioned in the last paragraph. Following this convention, a typed set may have the form:

$$\{(b \vdots \beta), (c \vdots \gamma)\} \vdots \Psi.$$

Here, the set has two members, of type β and γ. The set is of type Ψ. The operator $\vdots$, vertical ellipse, binds a type to its respective object.

2.2 Formation Rules

A set of basic types is assumed. Every member of the basic set is associated with a set in the domain. New sets can be formed out of existing ones using set operations. The type of the resultant set is determined as follows:

Let τ be the set of basic types. Then, T is the smallest set such that:

i) $\tau \subseteq T$.
ii) if X and Y $\in$ T then (X op Y) $\in$ T.

Here *op* stands for either ⊕, the type union operator, or : , the type intersection operator, defined below. They are used along with the union and intersection operations.

The set of basic types, τ, also contains three special basic types: the top, the bottom and the hash. The top type, T, is the most general type. It can be thought of as the disjunction of all types. Its polar opposite the bottom type, ⊥, or the absurd type, is the conjunction of all types. The empty set, then, can uniquely be identified as having the bottom type, that is, $\emptyset \vdots \bot$. It follows from this that there are many other empty sets which belong to different types or a combination thereof. A member can be referred to as having a top type. This is the extensional sense we are accustomed to from conventional sets. The ZF set can, then, be thought as having the top type, along with all its members. The hash, #, is the vacuous type, see §4 for an example.

2.3 Operations on Sets

Operations involving typed sets are similar to those involving ZF sets with the exception of taking types into consideration. In what follows and throughout, 'T' and '⊥' designate the top and bottom types, respectively.

(**Objects identity**) Let ≡ stands for identity between objects. For any set members x = (e1 ⁝ t1) and y = (e2 ⁝ t2),

$$x \equiv y \Leftrightarrow e1 - e2 \wedge [(t1 = t2)].$$

(**Type Casting/Shifting**) Typed set introduces a type-casting operation, which has no equivalent in conventional set. It has the following form: (X) (e ⁝ y). After this operation is carried out, the object denoted by 'e' exhibits the type specified by the type denoted by 'X', i,e., the operation returns the typed object denoted by (e ⁝ x).

In the following definitions, X and Y are any sets of types T1 and T2, respectively.

(Subset) $X \subseteq Y \Leftrightarrow \forall e\,[\,(e \in X \Rightarrow e \in Y)] \wedge (T1 = T2).$

(Set equality) $X = Y \Leftrightarrow [(X \subseteq Y \wedge Y \subseteq X) \wedge (T1 = T2)].$

(Union) $X \cup Y = \{x\,|x \in X \vee x \in Y\} \vdots (T1 \oplus T2).$

(Intersection)

a) $X \cap Y = \{x|x \in X \wedge x \in Y\} \vdots\ (T1 : T2).$

b) For any x and y,

$[(x \in X \wedge y \in Y) \wedge (x = (e \vdots T)\ y = (e \vdots T))] \Rightarrow [(e \vdots T) \in (X \cap Y) = Z \vdots (T \oplus T1)].$

(Properties of the operators : and $\oplus$) For any distinct types $T1 \neq \#$ and $T2 \neq \#$,

a) $[T1 \oplus T2] = [T2 \oplus T1].$ b) $[T1 \oplus T1] = T1.$ c) $[T \oplus T1] = T.$

d) $[\perp \oplus T1] = T1.$ e) $[T1 : T2] \neq [T2 : T1]$, where $T1 \neq \perp$ and $T2 \neq \perp$.

f) $[T1 : T1] = T1.$ g) $[T1 : \perp] = \perp.$ h) $[T : T1] = [T1 : T] = T1.$

Also, (i) and (j) follow from the above rules, and k-m hold for the vacuous type, #:

i) $[\perp \oplus T] = T.$ j) $[\perp : T] = \perp.$ k) $[\# : T1] = [T1 : \#] = T1.$

l) $[\# \oplus T1] = [T1 \oplus \#] = T1.$ m) $(x \vdots \#) = (x \vdots T1)$, for any type x.

The above identities show that the union operation within typed sets is associative and commutative. The intersection is neither, aside from the trivial cases. This can be attributed to the nature of union as a *par excellence* aggregate operation.

2.4 Predication and Graded Membership

One of the motivations for typed sets is to better accommodate predication in areas where the classical set is seen to be too limiting. In other words, the boundaries of the classical set need to be further "pushed" to more accurately represent what we conceive as an aggregate of disparate objects whose admission to the aggregate need not be definite. This involves allowing members of different membership levels. Thus, in our view, predication is a special case of set membership: members of a non-empty denotation of predicate both belong to the aggregate and participate in the same type of the predicate. Example: let T be the denotation of a predicate term, say *F*, with members as follows:

$$T = \{(o \vdots \alpha), (e \vdots \alpha'), (b \vdots \omega), (r \vdots \omega'), (c \vdots \gamma), (h \vdots \pi), (t \vdots \top)\} \vdots \Omega.$$

Although T has seven members, *F* can only be predicated of three members: *b*, *r*, and *t*. This is because the first two participate in the type Ω and *t* is member in some absolute sense – more on this kind of membership is below. The members *o*, *e*, *c* and *h* belong to the aggregate but *F* cannot be predicated of any of them. This is indicated by neither of them having an Ω type, i.e., those syntactically formed by ω or ω followed by prime(s). Notionally, the significance of this distinction is to capture graded membership: some objects fall within the concept completely while others fall only partially. An example of the latter is being an observer in a committee or a group – as was Russia's relation to the G-7 before being a full member; or, describing someone as allegedly being a member of a gang. Also, it is possible that a member is participating in the aggregate in a different capacity. This distinction plays an important inferential role. For if the objects *o* and *b* are the referents of some terms, say *a* and *c*, respectively, then *F(c)* is true while *F(a)* is false.

2.5 Quantification

Classical sets fit well with classical logic in terms of types assumed. Since a term in these systems is logical *par excellence*; only the referent of the term plays a semantic role and nothing else beside the referent. So, the universal quantifier is rephrased for *anything* or *everything* and similarly the existential goes by the colloquial *at least something* is such and such. Here, the type is implicitly the top type, T. With typed sets in mind, we are in agreement with the classical theorist: the domain contains definite things. But we differ in insisting that members of a set must not necessarily be of top type. For the way the result of isolating part of the domain must be reflected in the resulting set. That is, the resulting set consists of objects from the domain as well as the way they are perceived. Here we live up to Cantor's words of "a collection into a whole of distinct objects of *our* intuition or thought."[1] Thus, in typed sets, although we speak of type as part of the member, it is meaningless to speak of typed objects of the domain of quantification. For there are indefinite ways the objects of the domain can be designated and, as we have seen, typed sets may contain objects of different types. Consequently, the formulae of our language must specify the type under quantification. For instance, the universal quantifier must have the form:

$$[(\forall x)\, P(x) \Rightarrow R(x)]_{T}.$$

Here, the subscript indicates the type under quantification and the square brackets delineate the scope of the type t – thus, t (by default) is the top type, T.

Thus, we have two kind scopes: that of quantification and that of type. These scopes may not necessarily coincide. For instance the discourse "A cat is on the

1 Emphasis added.

mat. It is black." can be rendered, assuming the definite description is a referring term:

$$(\exists x)[\mathrm{cat}(x) \wedge \mathrm{on}(x, \mathrm{the\_mat}) \wedge \mathrm{black}(x')].$$

Here, the anaphor is not c-commanded by the antecedent. So it cannot be bound by the quantifier. The anaphor and the antecedent can still be type-bound – indicated by the prime in the argument of the predicate **black**. For type-binding needs not adhere to the c-command restriction. What matters here is the interpretation: it states that x' refers to a type (or mode of presentation, see §4) which can be used to determine the referent, denoted by *x*, for the predicate. This can be done since *x* and *x'* share the same unique type.

3 A Moment of Reflection

We briefly pause to address some concerns which might be raised. These include: 1) Are typed sets really necessary – after all classical set theory is a framework for all of mathematics; pairs or any other structure would do just fine? 2) Don't types violate ontological parsimony? 3) Is it still appropriate to call a typed set a "set"? In other words, aren't typed sets at odds with, if not running in the face of, the extensionalist's program in which the classical set plays a central role? We take up these issues in turn.

Prima facie it seems harmless to render a typed set member as a pair, say <a, b>. However, according to the most commonly accepted definition of a pair, this is just a short for the set {a, {a, b}}. But this would treat types as having an equal ontological status to and as being independent of their bearers. Since a particular's redness cannot be isolated from the bearing particular, similarly a type cannot be separated from its bearer. When we say a member may bear many types, we do not mean an object can acquire and shed its types at will. For types are intrinsically dependent on their bearers. This, however, doesn't mean that typed sets cannot be represented as pairs. They can, as long as the conceptual framework is taken into consideration. Calculus without the conceptual leap of the derivative would remain the old algebra. But once the underlying notion of the derivative was fully appreciated, the algebraic machinery was called upon to carry it out. Similarly, pairs, or any other mathematical structure, can be used as long as the underlying notion of typed sets is honored.

As for the ontological question, first, as argued above, types are not extraneous entities. The classical set implicitly assumes them. This is true not only of Russell's Type Theory and theories which assume urelements. Types are, also, present in ZF-axiomatic set theory. It might be argued that in the latter types exist only pre-theoretically as a means of enabling cumulative (or piece-wise) construction of sets. Admittedly, our use of types appears to differ from these cases. However, we argue that our way of utilizing types is a natural way to extend the notion of set where not only the aggregate's size is of interest to us but also the fabric of its composition – *ad hoc* vs natural totalities, Bittner and Smith (2003). Second, encoding the type in the member has merits: it blurs the boundaries between type hierarchies; the advantage of this can be appreciated if we consider how predication in Russell's theory of types (and similar ones) is severely curtailed in order to avoid paradoxes.

We conclude the ontological question by alluding to urelements. Axiomatic set theory assumes nothing but the empty set, everything else is built up bottom-up from the empty set. This rendition is seen as attractive by reductionists – mathematical or ontological – but counterproductive and unnecessary by others, e.g., Simon (2005) and Benacerraf (1965). Also, Ionicioiu (2013) questions the ability of the axiomatic set theory to be a framework for Quantum Mechanics. Ionicioiu discusses alternatives to rectify the situation. One of these is to postulate distinct urelements.

Thus, assuming the empty set as the sole inhabitant is not a decisive factor or a benchmark everyone must have to live up to. In fact, our postulating of "divisible" urelements may turn out to have an ontological advantage: it is possible that what we termed a "core" can also be further composed of a core and a type. This is crucial for middle-level ontological theorizing where a classical-set-based ontology would insist on specifying the ultimate small building blocks beforehand, Bittner and Smith (*ibid*).

Finally, it is reasonable to question where typed sets stand: are they still real "sets" or has the subject changed? We believe they are still sets. As mentioned, we think of typed sets as natural extension to classical set. Similar to classical set, a typed set has an identity criterion – where other notions such as properties lack – and operations on typed sets are pretty much the same as (the well understood) operations of classical set. This is of significant importance when one considers the advantage of (classical) sets over alternatives, e.g., property theory, as pointed out in Jespersen (2008). However, there is one more feature typed sets have: every set is unique and, thus, need not be named. This is similar to the advantage lambda notation has over the conventional function notation. The merit of this feature can be utilized in computing the meaning of multiple modifiers of NL expressions where indefinite number of intermediary sets may ensue. The extensional character of set, we argue, is not undermined by the introduction of types. In fact, it is the opposite: non-extensional aspects of NL can be subjected to the extensional machinery via typed sets, as shown below.

4 Application

In this section we show how the meaning of some NL constructs can be represented and computed using typed sets. In this demonstration, we assume Russellian (or singular) propositions as well as the direct theory of reference of proper names. Our choice is influenced by the appeal of these notions as well as the challenges they face in the treatment of belief contexts and fictional objects.

We assume a model containing a set of the objects of the domain and a set of types – for every construct of the object language with the exception of connectors and proper names. We like to think of these types as a Fregean Mode of Presentation, but not Fregean Senses, as utilized in Abdullah and Frost (2013). Computation is carried out compositionally following the rules of §2 and function application.

4.1 Modification

Adjective-noun combinations are believed to form a semantically-motivated 3-level-hierarchy, Kamp and Partee (1995). Conceived as such, these compounds seem to challenge the syntax-semantics isomorphic doctrine. The accepted uniform treatment rests on PW semantics.

Consider, a) *That is a red car* and b) *Alice is a good pianist*. If the denotations of both red and car are taken as sets, then (a) is true if, and only if, the referent of *that* belongs to the intersection of the two sets. But (b) has quite different truth-conditions. The adjective *good* in (b) is said to have two possible readings: reference-modifying and referent-modifying. The former pertains to Alice herself, that is, her character. The latter, pertains to her piano performance. The trouble starts when an at-face-value ZF set-theoretic treatment is given to *good* similar to that of red in (a). For Alice may happen to be a teacher too. This would wrongly lead to the conclusion that Alice is a good teacher which doesn't follow from (b) and the sentence *Alice is a teacher*. With the distinction between set-membership and predication in mind, type-endowed sets can correctly and uniformly encode the truth-conditions of the adjective hierarchy. To demonstrate, consider (b), again, and c) *Alice is a teacher*. In our account, common nouns and all adjectives denote (typed) sets. So, the facts from (b) and (c) are represented as follows, where the symbol '||||' designates the denotation function:

$\|Alice\| = (e \vdots \#)$[2] or just e; $\|good\| = \{(e \vdots \gamma), (e \vdots T), ..\} \vdots A$;
$\|teacher\| = \{(e \vdots \beta), ..\} \vdots B$; $\|pianist\| = \{(e \vdots \gamma), .. \} \vdots \Gamma$.

It should be obvious that the (wrong) inference *Alice is a good teacher* cannot be reached. This is because the member $(e : \beta)$ does not belong to the intersection of the denotations of *teacher* and *good*, due to type mismatch. Incidentally, it should be noted that Alice is a member of the set ||good||in two different capacities. It follows by rule Intersection-(b) of §2.3 that the assertion Alice is a teacher who is good holds. This is because $(e \vdots T)$ is in the intersection of the denotations of *teacher* and *good* and the resultant set is of type T. Thus, polysemous words such as *good* can be accommodated without postulating multiple lexical entries, Cf Siegel (1976).

Adjectives of the third level are characterized by having referents that neither belong to the denotations of the intersection of the adjective and the common noun nor to a subset of the denotation of the common noun, (d) and (e) are examples: d) *That is a fake gun* and e) *Lefty is an alleged murderer*. When *fake* and *gun* are assigned the types A and B, respectively, (d) and (e) can be rendered as follows:

$\|that\| = g$; $\|fake\| = \{(g \vdots \alpha), ..\} \vdots A; |$ $|gun\| = \{(g \vdots \alpha), ..\} \vdots B$.

We immediately see that we can predicate fake of g but not gun. Thus, we can say about the object g that it is fake – for it is in ||fake||and has the same type as the set – but we cannot describe it as a gun. (e) can be rendered as follows

$\|Lefty\| = l; \|$ $alleged\| = \{(l \vdots \psi), ..\} \vdots \Omega$; $\|murderer\| = \{(l \vdots \omega), ..\} \vdots \Psi$.

Here, Lefty can neither be predicated of alleged nor murderer. But the statement Lefty is alleged murderer can be asserted as follows, where F is a function from types and sets to sets and T is a function from sets to types:

$$\|Lefty\| \in [F\ (T(\|murderer\|), \|alleged\|) \cap \|murderer\|].$$

The extra work here is needed since, as things stand, the intersection of the denotations of *alleged* and murderer is empty. To get the correct reading, we need to do two things: to extract members of type ||murderer||from the set ||alleged||and to form a new set with members of type ||alleged||. This is what the function F is

2 Here a Millian view of proper names is assumed.

purported to do. Then intersection can be carried out. The whole expression, then, can be evaluated to True if, and only if, ||Lefty||is in the intersection.

Noun-noun combinations can be handled in a similar way. Expressions such as *child murderer* are ambiguous – describing someone who murders children or a child who murders. The inclusive *or* reading of *Sparky is a child murderer* can be rendered as:

$$||\text{Sparky}|| = s; \;|\quad |\text{child}|| = \{(s \vdots \sigma), (s \vdots \psi), ..\} \vdots \Sigma; \quad ||\text{notorious}|| = \{(s \vdots \psi), ..\} \vdots X;$$
$$||\text{murderer}|| = \{(s \vdots \sigma), (s \vdots \psi), ..\} \vdots \Psi.$$

If it happens that Sparky is a notorious murderer then the assertion Sparky is notorious child murderer is derivable since $(s \vdots \psi)$ belongs to the intersection of the three sets. It should be noted that given membership-predication distinction the assertion Sparky is notorious is semantically ill-formed as does Lefty is alleged. This is reflected in the representation by having both Spark and Lefty as members but not type-participants in the respective sets, making predication unattainable.

4.2 Non-Existents

Non-existents, fictional characters or empty names, as they are sometimes called, have had an uneasy "existence" in philosophy. However, there is an increasing interest in them, e.g., Parsons (1975), Routley (1980), Priest (2005), Rapaport (2013)), and we also, as fictional entities are important part of discourse. Here, we are interested in providing a viable semantic account of fictional entities. Our goal is to keep the semantic and ontological tracks separate by means of our distinction between set-membership and predication and, without appealing to notions such as *story operator*, *pretense* or PWs. The latter seems to be the favorite choice for many authors. Some authors, e.g., Priest (2005), even invokes Impossible Worlds (IWs) to accommodate non-existents. Although Priest's approach is elaborate and semantically elegant, we find the notion of IWs problematic. For if the laws of logic can break down in some IW one wonders what else is lost. Priest thinks the sentence *x is round and x is square* holds in some IW this is because, according to him, such a world is impossible. But, then one can add *x has counterparts in some other worlds*. Isn't this reminiscent of Russell's objection to Meinong? Of course, such an observation can hardly escape Priest's notice. But we are still left with many questions: Which relations/laws do hold and which do not? Does the reference relation hold "there"? In short, one wonders how these "logical black holes" can be tamed.

To properly address fictional entities (and empty names in general), we observe that: i) expressions such as *fictional characters* must be accounted for, as they are undeniably meaningful; and, ii) The act of naming a member of such an extension is *deliberate* and conscious – maybe with some exceptions, e.g., cases of deities. This act involves articulating the properties of such named entities. This very act amounts to determining a (Fregean) sense. Therefore, a sentence such as *Santa is a man* can semantically be accounted for as follows:

$$||\text{Santa}|| = (s \vdots \bot); \qquad ||\text{man}|| = \{(s \vdots \alpha), ..\} \vdots A.$$

Here, the denotation of 'Santa' is ascribed the absurd type upon naming. The significance of this ascription and representation is that existential generalization does not apply. Thus, the inference:

$$\text{man(Santa)} \vdash [(\exists x)\ \text{man}(x)\ ]_{\top},$$

is invalid since the referent of 'Santa' is not within the domain of quantification. To validate such an inference the quantification scope needs to be explicitly specified as of type $\bot$, as per §2.5. This is harmless in that it incurs no ontological consequence. In fact it is desirable, e.g., in hypothesizing and considering alternative scenarios.

Advantages of this analysis include: i) individuation is possible as only named entities are admitted into discourse; ii) the ontology is kept non-bloated – as only named entities are admitted; and, iii) fictional entities are not isolated from the general discourse, as in other approaches such as pretense and story operator. For the interplay between discourses is the norm, e.g., *Seinfeld played Seinfeld; the latter is liked by his friends but not the former*, *John admires all detectives*, etc.

4.3 Non-Extensional Contexts

Our approach to belief contexts builds on Russell's well-known scope approach, McKay (2010). An attractive feature of this approach is that it has the potential of adhering to an isomorphism between the syntax and the semantics: every syntactically *de re* is semantically *de re*, and the same goes for *de dicto*. For example, a sentence of the form *John believes someone is a spy* is *de re/de dicto* ambiguous and the ambiguity can be accounted for by appealing to the scope of both the belief verb and the quantifier. This simple solution was challenged by Quine (1965) on the grounds that the de re reading involves quantifying into intensional contexts, which renders the reading inconsistent. We believe that by using typed sets and by appealing to type-scope, Quine's challenge can be met without postulating lexical ambiguity.

As pointed out in §2.5, it is necessary to determine the type of an object as well as the scope where such an object is to be identified with the associated type. For assuming just one such a unique type leads to inconsistencies, as observed by Quine. That is, in our typed set notation, this can be expressed as: $\exists x[\mathrm{bel}(j, \mathrm{spy}(x))]_{\top}$.

However, as Recanati (2000) rightly observes, there is an ambiguity which stems from the transparent and opaque reading which is "indeed orthogonal to, the relational/notional ambiguity." Within typed sets, this can easily be explained as type scope phenomenon. In our notation, this can be brought out as in (a) and (b):

a) $[(\exists x)\mathrm{bel}(j,\mathrm{spy}(x))]_{\top}$.   b) $[(\exists x)\mathrm{bel}(j,\mathrm{spy}(x))]_{\Psi}$.

That is, although the reading is de re, two distinct type scopes are involved, $\top$ and the belief-induced type Ψ. (a) accounts for the transparent and (b) for the opaque. Thus, we end up with two distinct propositions, involving different guises, aka, types. These ambiguities, along with the *de dicto*, can be represented, provided there are spies actually, as:

$$||\mathrm{believe}||= \{(j,<(e \vdots \bot),SPY>),(j,<(e \vdots \Psi),SPY>);\quad (j,||\lambda\exists(x)\,[\mathrm{bel}(j,\mathrm{spy}(x))]||),\ldots\} \vdots Y.$$

Here **belief** is uniformly treated as a relation between an object and a proposition of the form <object, S>, which is true iff the object named is a member of the set S. Our use of λ-operator follows Zalta (2010) in naming propositions.

Transitive intensional verbs can be treated similarly. For example, *John seeks a unicorn* is ambiguous between the specific and unspecific readings, which can be represented as:

$$||\mathrm{seek}||= \{(j,||\lambda\exists(\Sigma)x\,||),\ldots\} \vdots T,$$

where (Σ) x is a type-casting operation and the type Σ represents the type associated with the set of unicorns. Again, the lambda operator indicates a name, this time of an entity. This can be generalized for other quantifiers. Here, the pair is understood as relating John to some anonymous entity, pretty much as *John seeks Santa*.

5 Conclusion

We have introduced typed sets arguing that they are a natural extension to classical sets. We argued that a finer granularity of analysis is gained by regarding an urelement as being made of a core and a type. We showed that this conception helps avoid the conflation of predication and aggregation. We argued that the former is special case of the latter. We showed that operations on typed sets are very similar to operations involving conventional sets. Next, we explained how quantification must be understood within typed sets, insisting that the domain of quantification contains just objects and language formulae must specify the type under which quantification takes place. We demonstrated the utility of typed sets by providing a uniform treatment of NL expressions which are traditionally considered to be non-extensional, without resorting to any other notions and requiring no more resources than those available to first-order sets.

Bibliography

Abdullah, N. and Frost, R. (2013) Senses, Sets and Representation. In *The Proceedings of the 2013 Meeting of the International Association for Computing and Philosophy*: http://www.iacap.org/proceedings_IACAP13/paper_25.pdf.Accessed 5/4/2014.
Abbott, B. (2010) *Reference*. Oxford University Press.
Benacerraf, P. (1965) What Numbers Could Not Be. *Philosophical Review* 74, 57–73.
Bittner, T. and Smith, B. (2002) A theory of Granular Partitions. In Worboys, M. and Duckham, M. (eds.), *Foundations of Geographic Information Science*.
Bombieri, Enrico. (2010) The Mathematical Truth.. IAS. Princeton. (Lecture).
Carnap, R. (1948) Meaning and Necessity. University of Chicago Press.
Chierchia, G. and Turner, R. (1988) Semantics and Property Theory. *Linguistics and Philosophy*, 11:261–302.
Cocchiarella, N. (1988) Predication versus Membership in the Distinction between Logic as Language and Logic as Calculus. *Synthese*, Vol. 77, No. 1, Logic as Language, pp. 37–72.
Hirst, G. (1988) Semantic interpretation and ambiguity. *Artificial Intelligence*. 34, 131–177.
Ionicioiu, R. (2012) Is Classical Set theory Compatible with Quantum Experiments? Premier Institute for Theoretical Physics, Waterloo, Canada. (Lecture)
Jespersen, B. (2008) Predication and Extensionalism. *Journal of Philosophical Logic*. Vol. 37, No. 5:479–499.
Kamp, H. and Partee, B. (1995) Prototype Theory and Compositionality. *Cognition* 57:129–191.
Loux, M. (2006) Metaphysics: a contemporary introduction. 3rd ed. N.Y.: Routledge.
McGinn, C. (2000) Logical Properties: Identity, Existence, Predication, Necessity, and Truth. Oxford University Press.
McKay, T. and Nelson, M. (2012) Propositional Attitude Reports. In *The Stanford Encyclopedia of Philosophy (Winter 2010 Edition)*, Edward N. Zalta (ed.), URL = <http://plato.stanford.edu/archives/win2010/entries/prop-attitude-reports/>.
Parsons, T. (1975) A Meinongian Analysis of Fictional Objects. *Grazer Philosophische Studien*, 1:73–86.
Partee, B. (1979) Semantics – mathematics or psychology? In Bäuerle, R., Egli, U. & Stechow A. von (eds.), *Semantics from Different Points of View*, 1–14. Berlin: S–V.
Priest, G (2005) *Towards Non-Being: The Logic and Metaphysics of Intentionality*. Clarendon Press.

Rapaport, W. (2013) Meinongian Semantics and Artificial Intelligence, in special issue on "Meinong Strikes Again: Meinong's Ontology in the Current Philosophical Debate", (eds.), Mari, L. & Paoletti, P., *Humana Mente: Journal of Philosophical Studies*.
Recanati, F. (2000) Relational Belief Reports. *Philosophical Studies* 100 (3):255–272.
Routley, R. (1980) Exploring Meinong's Jungle and Beyond. *RSSS*, Australian National University, Canberra.
Siegel, M. (1976) *Capturing the adjective*, University of Massachusetts.
Quine, W.V. (1956) Quantifiers and propositional attitudes. *Journal of Philosophy* 53, 177–187.
Quine, W.V. (1948) On What There Is. *Review of Metaphysics*. 2(5):21–38.
Woods, W. (1985) What's in a Link: Foundations for Semantic Networks. In Brachman, R. and Levesque, H. (1985) (eds.), *Readings in Knowledge Representation*. pp. 217–241.
Zalta, E. (2010) Logic and Metaphysics. *Journal of Indian Council of Philosophical Research* 27 (2):155–184.

Ahmed Magdy Ezzeldin, Yasser El-Sonbaty, and Mohamed Hamed Kholief

Exploring the Effects of Root Expansion, Sentence Splitting and Ontology on Arabic Answer Selection

Abstract: Question answering systems generally, and Arabic systems are no exception, hit an upper bound of performance due to the propagation of error in their pipeline. This increases the significance of answer selection systems as they enhance the certainty and accuracy of question answering. Very few works tackled the Arabic answer selection problem, and they did not demonstrate encouraging performance because they use the same question answering pipeline without any changes to satisfy the requirements of answer selection. In this paper, we present "ALQASIM 2.0", which uses a new approach to Arabic answer selection. It analyzes the reading test documents instead of the questions, utilizes sentence splitting, root expansion, and semantic expansion using an automatically generated ontology. Our experiments are conducted on the test-set provided by CLEF 2012 through the task of QA4MRE. This approach leads to a promising performance of 0.36 accuracy and 0.42 c@1.

1 Introduction

Question Answering for Machine Reading Evaluation (QA4MRE) is a kind of question answering (QA) that is meant to evaluate the ability of computers to understand natural language text by answering multiple choice questions, according to the certainty of each answer choice, inferred from a reading passage. Thus, QA4MRE depends mainly on answer selection [1] as passage retrieval is not required.

In 2005, it was noticed in CLEF (Conference and Labs of the Evaluation Forum) that the greatest accuracy reached by the different QA systems was about 60%, where on the other hand 80% of the questions were answered by at least

Ahmed Magdy Ezzeldin, Yasser El-Sonbaty, Mohamed Hamed Kholief: College of Computing and Information Technology, AASTMT Alexandria, Egypt, e-mail: a.magdy@a1works.com, {yasser, kholief}@aast.edu

one participating system. This is due to the error that propagates through the QA pipeline layers: (i) Question Analysis, (ii) Passage Retrieval, and (iii) Answer Extraction. This led to the introduction of the Answer Validation Exercise (AVE) pilot task, in which systems were required to focus on answer selection and validation, leaving answer generation aside. However, all the QA systems from CLEF 2006 to 2010 used the traditional Information Retrieval (IR) based techniques and hit the same accuracy upper bound of 60%. By 2011, it was a must to devise a new approach to QA evaluation that forces the participating systems to focus on answer selection and validation, instead of focusing on passage retrieval. This was achieved by answering questions from one document only. This new approach, which was named QA4MRE, skips the answer generation tasks of QA, and focuses only on the answer selection and validation subtasks [2]. The questions used in this evaluation are provided with five answer choices each, and the role of the participating systems is to select the most appropriate answer choice. The QA4MRE systems may leave some questions unanswered, in case they estimate that the answer is not certain. C@1 was also introduced by Peñas et al. in CLEF 2011, which is a metric that gives partial credit for systems that leave some unanswered questions in case of uncertainty [3]. Arabic QA4MRE was then introduced for the first time in CLEF 2012 by two Arabic systems, which will be discussed in section 2.

2 Related Works

A few works tackled the answer selection task in QA and most of these works used redundancy IR based approaches and dependence on external sources like Wikipedia or different Internet search engines. For example, Ko et al. used Google search engine and Wikipedia to score answer choices according to redundancy, using a corpus of 1760 factoid questions from Text Retrieval Conference (TREC) [4]. Mendes and Coheur explored the effects of different semantic relations between answer choices and the performance of a redundancy-based QA system, also using the corpus of factoid questions provided by TREC [5]. However, these approaches proved to be more effective in factoid questions, where the question asks about a date/time, a number or a named entity, which could be searched easily on the Internet and aligned with the answer choices. On the other hand, QA4MRE supports the answer choices of each question with a document and depends on language understanding to select the correct answer choice, which can generalize for other kinds of questions like causal, method, list and purpose questions.

In CLEF 2012, Arabic QA4MRE was introduced for the first time. Two Arabic systems participated in this campaign. The first system created by Abouenour et al. [6] (IDRAAQ), has a 0.13 accuracy and a 0.21 c@1. It uses JIRS for passage retrieval, which uses the Distance Density N-gram Model, and semantic expansion using Arabic WordNet (AWN). IDRAAQ does not use the CLEF background collections. Its dependence on the traditional QA pipeline, to tackle QA4MRE, is the main reason behind its poor performance.

The second system by Trigui et al. [7] depends on an IR based approach and achieves the accuracy and c@1 of 0.19. It collects the passages that have the question keywords and aligns them with the answer choices, then searches for the best answer choice in the retrieved passages. It then employs semantic expansion, using some inference rules on the background collections, to expand the answer choices that could not be found in the retrieved passages. However, these two Arabic systems do not compare to the best performing system in the same campaign, which was created by Bhaskar et al. [8]. It uses the English test-set and performs at an accuracy of 0.53 and c@1 of 0.65. This system combines each answer choice with the question in a hypothesis and searches for the hypothesis keywords in the document, then ranks the retrieved passages according to textual entailment. Thus, the English system outperforms the other two Arabic systems mainly because it depends on analyzing the reading test documents and uses textual entailment.

In CLEF 2013, we introduced ALQASIM 1.0, which is an Arabic QA4MRE system that analyzes the reading test documents, where answer choices are scored according to the occurrence of their keywords, their keywords weights, and the keywords distance from the question keywords. It performs at an accuracy of 0.31 and c@1 0.36 without using CLEF background collections [9].

In the next section, we will present the architecture of our new Arabic QA4MRE system ALQASIM 2.0. Then in section 4, we will demonstrate its performance and compare it with the two Arabic QA4MRE systems at CLEF 2012.

3 Proposed System ALQASIM 2.0

In this section, ALQASIM 2.0 components will be illustrated, and then the integration of these components will be explained in detail. The system consists of three main modules as shown in Fig. 1: (i) Document Analysis, (ii) Question Analysis, and (iii) Answer Selection. The document analysis module applies morphological analysis, sentence splitting, semantic, root and numeric expansion to the reading test document and saves the document keywords and their expansions to an inverted index. The question analysis module applies morphological analysis, and identifies some question patterns to highlight the question focus, then searches the inverted index for the question keywords and returns the question snippets. The answer selection module applies morphological analysis to the answer choices to extract their keywords, then applies a constrained scored search in proximity to the found question snippets from the inverted index and selects the best scoring answer choices that it finds if any.

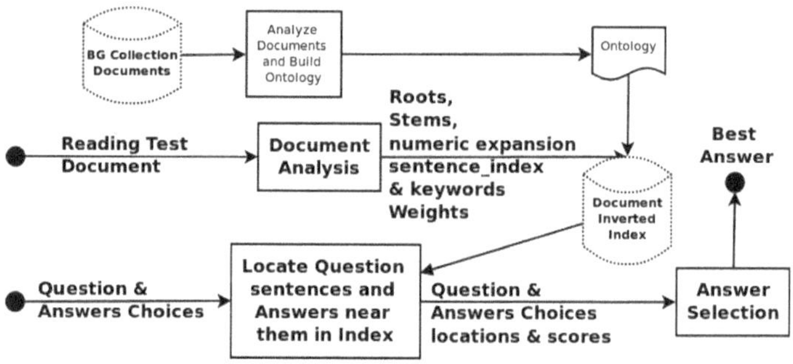

Fig. 1: ALQASIM 2.0 Architecture.

3.1 Document Analysis

The document analysis module is crucial for the answer selection process. It gives context for morphological analysis, and makes relating the question snippet to the answer choice snippet easier. In this module, the document is split into sentences, and the sentences indexes are saved in an inverted index for future reference. The text words are then tokenized, stemmed, Part-of-Speech (PoS) tagged, and stop words are removed. Words roots are then extracted, and textual numeric expressions are rewritten in digits. The word, light stem, root, and numeric expansion are all saved in the inverted index that marks the location (index) of each word occurrence. The words are also expanded using an ontology of hypernyms and their hyponyms extracted from the background collections provided by CLEF 2012. In the coming subsections, the components of the document analysis module are explained and discussed.

3.1.1 Inverted Index

The inverted index is an in-memory hash map data structure, created for each document. The key of this hash map is a string token that could be a word, stem, root, semantic expansion, or digit representation of a numeric expression. The value is a list of the locations of that token in the document. The inverted index also con-

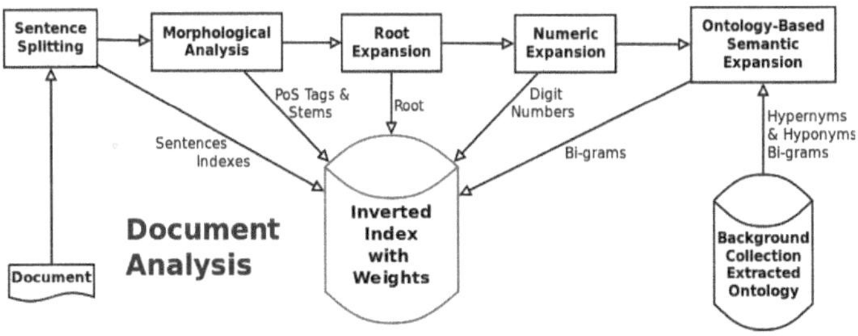

Fig. 2: Document Analysis module architecture.

tains the weight of each token, which is determined according to its PoS tag and whether it is a named entity (NE) as detailed in Table 1. The intuition behind setting tokens weights is to give higher weights for tokens that are more informative. NEs mark a question or an answer snippet easily, thus they are given the highest weight. Nouns convey most of the meaning as Arabic sentences are often nominal. Adjectives and adverbs help to convey meaning as they qualify nouns and verbs, thus they are given high weights but not as high as NEs and nouns. Verbs have a lower weight as they are less informative. Prepositions, conjunctions and interjections were already removed in the stop words removal process before saving the words to the inverted index.

Tab. 1: Weights determined according to PoS tags and NEs.

PoS Tag / Named Entity	Weight
Named Entity (person, location, organization)	300
Nouns	100
Adjectives and Adverbs	60
Verbs	40
Other PoS Tags	20
Prepositions, conjunctions, interjections and stop words	N/A

3.1.2 Morphological Analysis Module

The morphological analysis module is the most important one. MADA is used in this module to provide tokenization, PoS tagging, light stemming, and number determination of each word. MADA is an open source morphological analysis toolkit, created by Habash et al. [10]. It takes the current word context into consideration, as it examines all possible analyses for each word, and selects the analysis that matches the current context using Support Vector Machine (SVM) classification for 19 weighted morphological features. MADA has an accuracy of 86% in predicting full diacritization. The morphological analysis module also saves the light stem of each word in the search inverted index, with a weight that is determined by the word PoS tag as mentioned in the previous subsection.

3.1.3 Sentence Splitting Module

The sentence splitting module marks the end of each sentence by the period punctuation mark that is recognized by MADA PoS tagger. MADA helps the sentence splitting module by differentiating between the punctuation period mark and other similar marks, like the decimal point. The sentence splitting process is very important for locating question and answer snippets.

3.1.4 Root Expansion Module

Using a root stemmer to expand words proved to be effective in many NLP tasks on morphologically rich and complex languages like Arabic [11]. The root expansion module uses an Arabic root stemmer to extract each word root and save it to the inverted index as an expansion to the word itself. Three root stemmers have been tested with ALQASIM 2.0, which are: (i) Khoja [12], (ii) ISRI [13] and (iii) Tashaphyne[1] root stemmers.

3.1.5 Numeric Expansion Module

The numeric expansion module extracts numeric textual expressions, and saves their digit representation to the inverted index. It also takes word number tags, provided by the morphological analysis module, and adds the number 2 with each dual word. For example, the word "شهادتين" that means "2 Certificates" is saved to the inverted index as two words in the same location: "2" and "شهادة / certificate".

1 **Tashaphyne**, Arabic light Stemmer/segment :1)2) http://tashaphyne.sourceforge.net/

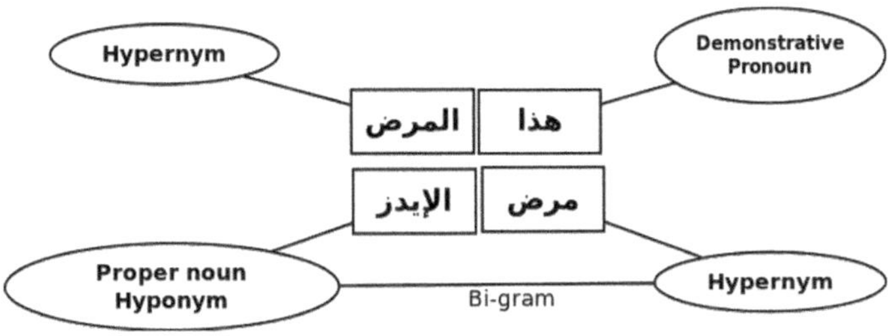

Fig. 3: The intuition behind extracting the background collections ontology.

3.1.6 Ontology-Based Semantic Expansion Module

ALQASIM 2.0 uses the background collections provided with the QA4MRE corpus from CLEF 2012, which is a group of documents related to the main topic of the reading test document. The ontology-based semantic expansion module expands any hypernym in the document with its hyponyms extracted from the background collections ontology. A hyponym has a "type-of" relationship with its hypernym. For example, in the bi-gram "مرض الإيدز / AIDS disease", the word "مرض / disease" is a hypernym and the word "الإيدز / AIDS" is a hyponym because AIDS is a type of disease. The ontology is extracted from the background documents collections according to two rules as illustrated in Fig. 3:

1. Collect all hypernyms from all documents by collecting the Demonstrative Pronouns (e.g. هذه، هذا، هؤلاء، ذلك، تلك، أولئك، أولائك) followed by the noun that starts with "ال / Al" determiner. This noun is marked as a hypernym. E.g. "هذا المرض / this disease"
2. Collect any hypernym of the extracted list followed by a proper noun, considering the proper noun as a hyponym. This relation in Arabic is named "Modaf"-"Modaf Elaih" relationship, which often denotes a hypernym-hyponym relationship, where the "Modaf" (the indeterminate noun) is the hypernym and the "Modaf Elaih" (the proper noun) is the hyponym. E.g. "مرض الإيدز / AIDS disease".

3.2 Question Analysis

The main purpose of the question analysis module is to extract the question keywords, search for them in the inverted index and return the found question snippets with their scores and sentences indexes. It applies tokenization, light-stemming, and PoS tagging on the question text. Stop words are removed from the question words, and the remaining keywords roots are then extracted, using

ISRI root stemmer, and added to the question keywords. Some question patterns are also identified and a boost factor is applied for some keywords in them. For example, the pattern "ما ال" or "What is the" is most of the times followed by a hypernym that is the focus of the question, so a boost multiplier is applied to the weight of that hypernym (focus) to mark its importance. The inverted index is then searched for the question keywords to find the question scored snippets and sentences indexes, which are used in the answer selection module. Question snippets are scored according to the number of keywords and their weights found in each snippet as shown in Equation 1.

$$Score = \left(\sum_{i=1}^{N} W_i\right) + (N - K). \quad (1)$$

Where:
- **Score:** score of a question or answer snippet
- **N:** number of found keywords
- **W_i:** weight of each keyword
- **K:** number of all keywords

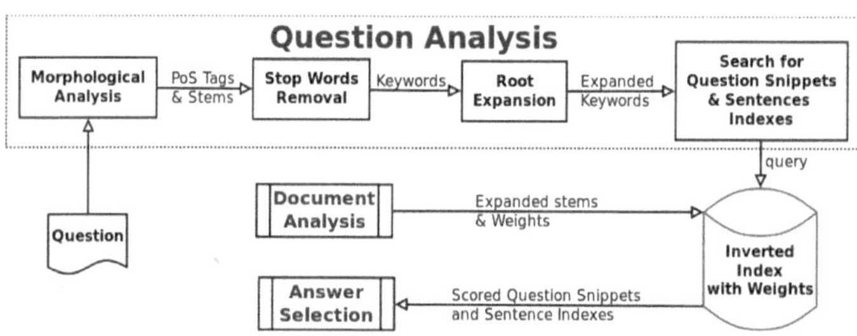

Fig. 4: Question Analysis module architecture.

3.3 Answer Selection

In the question analysis module, the best scoring question snippets and their sentences indexes are retrieved. In this module, each answer choice is analyzed exactly like the question, and the answer choice keywords are searched in the inverted index. The answer search is constrained to the sentence of the question snippet and the two sentences around it. The best scoring "found" answer choice is selected as the question answer. If there are no "found" answer choices, or the scores of the highest two or more answer choices are equal, then this means that the system is not sure which is the most appropriate answer choice, and marks the question as unanswered as shown in Fig. 5.

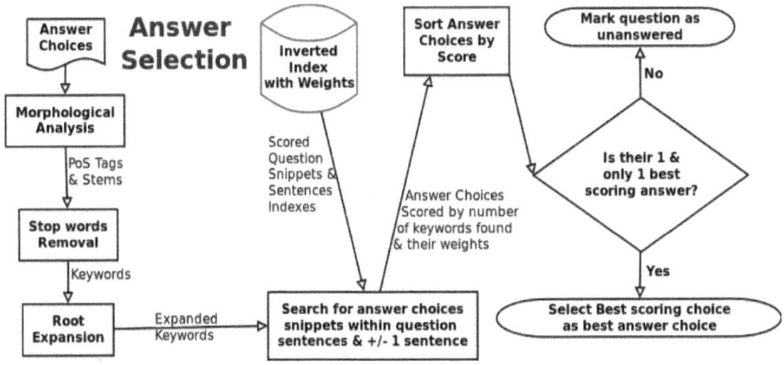

Fig. 5: Answer Selection module architecture

4 Performance Evaluation

In this section, the performance of ALQASIM 2.0 will be evaluated using a test-set provided by the QA4MRE task at CLEF 2012 [1]. The metrics used to evaluate the system are accuracy and c@1. Accuracy is the number of relevant items that are retrieved and the number of not relevant items that are not retrieved divided by the number of all items. C@1 was introduced by Peñas et al. in CLEF 2011 to give partial credit for systems that leave some unanswered questions in case of uncertainty [3]. The first run of the system uses the baseline approach, which consists of sentence splitting, light stemming, numeric expansion, weights defined according to PoS tags, and boost multipliers for some question patterns. The second run uses the baseline approach in addition to applying AWN semantic expansion. A dictionary of synonyms is generated from AWN to expand a word by adding its synonyms in the inverted index, and marking these synonyms with the same location of the original word. The third run applies semantic expansion using the background collections ontology to the baseline approach. The stem of each word in the document is expanded with its hyponyms if it is a hypernym, according to the explanation of the hypernym / hyponym relationship in section 3.1.6. The fourth run introduces root expansion using ISRI Arabic root stemmer (as explained in section 3.1.4), in addition to the baseline approach and the background collections ontology semantic expansion. Table 2 compares the results of these four runs.

In comparison to the other related Arabic QA4MRE systems reported in the literature, ALQASIM 2.0 shows superior performance in terms of accuracy and c@1 as shown in the chart in Fig. 6.

On comparing the performance of the three root stemmers (Khoja, ISRI and Tashaphyne), it is noticed that ISRI achieves the best performance in terms of ac-

Tab. 2: Performance of the 4 runs of ALQASIM 2.0.

Root Stemmer	Accuracy	C@1
Run (1) Baseline	0.29	0.38
Run (2) Baseline + AWN Semantic Expansion	0.28	0.36
Run (3) Baseline + Bg Ontology Expansion	0.32	0.40
Run (4) Baseline + Bg Ontology + Root Expansion	**0.36**	**0.42**

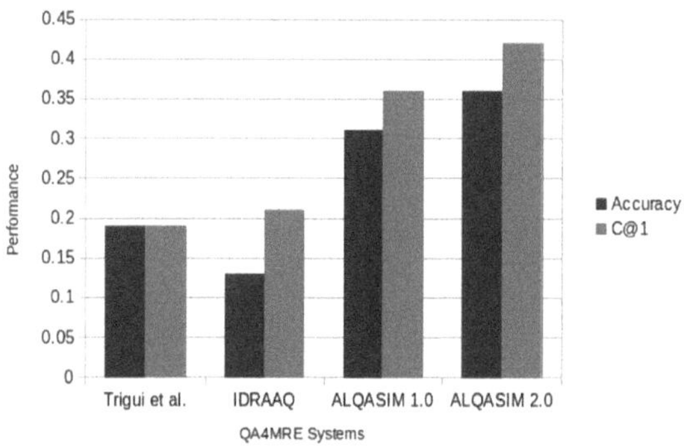

Fig. 6: Comparison between QA4MRE systems.

curacy and c@1. It is also noticed that Khoja stemmer does not improve performance and Tashaphyne root stemmer even degrades accuracy by 1% and c@1 by 4% as illustrated in Table 3.

Tab. 3: The effect of Khoja, ISRI and Tashaphyne stemmers on ALQASIM 2.0.

Root Stemmer	Accuracy	C@1
ALQASIM 2.0 without root expansion	0.32	0.4
ISRI	**0.36**	**0.42**
Khoja	0.32	0.4
Tashaphyne	0.31	0.36

5 Discussion and Future Work

From the results of Run (1), it is evident that using sentence splitting as a natural boundary for answer selection has a significant effect on performance. An accuracy of 0.29 and a c@1 of 0.38 demonstrate that the approach of sentence splitting and searching for the answer within proximity to the question sentence is efficient in pinpointing the correct answer choice, and even more efficient in ignoring wrong answer choices. This is due to the fact that most of the questions are answered either in the same sentence or within a distance of one sentence from the question sentence.

AWN semantic expansion degrades performance by about 1% to 2%. This is due to the generic nature of AWN that does not provide a specific context for any of the topics of the test-set. Thus, the question and answer choices keywords, that may be different in meaning, are expanded to be semantically similar, which makes it harder for the system to pinpoint the correct answer among all the false positives introduced by the semantic expansion process. For example, AWN expands only well-known words like "سياسة / policy or politics" and "نظام / system or arrangement" without any sense disambiguation according to context, and fails to expand context specific words like "الإيدز / AIDS" and "الزهايمر / Alzheimer". On the other hand, the automatically generated background collections ontology improves performance by about 3% and proves that it is more effective in the semantic expansion of document specific terms. For example, the ontology states that "Alzheimer" is a "disease", and that "AIDS" is a "disease", an "epidemic" and a "virus". The above-mentioned approach of building the ontology is most of the time capable of defining the right hypernym / hyponym pair, which helps to disambiguate many questions that ask about the hyponym while having the hypernym among the question keywords. This approach increases the overlap between the question and answer terms, and makes it possible to boost the weight of the hypernyms in some question patterns to imply their importance.

The last Run shows the effect of root expansion on answer validation. An improvement in performance of about 4% is achieved by using ISRI root stemmer. Thus, it is clear that root expansion makes it easier to pinpoint the correct answer especially when light stemming is not enough. For example, the question may ask about a noun, and the question snippet in the document has its verb which has a different derivational form. It is also clear that over stemming using Tashaphyne stemmer actually degrades performance as it gives the same root for some terms that are semantically different. For example, the word "سوءا / worse", "ماء / water", and "سواء / alike" have the same root in Tashaphyne when they are completely different in meaning. Khoja stemmer does not affect performance as it is not capable of stemming new words that are not found in its dictionary. Thus, in most cases, it either returns the original word or its light stem, which means no expansion is being carried out, taking into consideration that the light stem (provided by MADA) and the original word are already saved in the inverted index. When Khoja and ISRI root stemmers were tested on 15163 words from the above mentioned test-set, Khoja stemmer returned the roots of only 0.23%, the light stem (provided already by MADA) of 24.05% and the word without any stemming for 75.71% of the words.

On the other hand, ISRI stemmer returned the roots of 74.68%, light stem of 21.13% and the word without any stemming for only 4.18% of the words.

From the illustrated results, it is evident that the performance of ALQASIM 2.0 is superior to its Arabic competitors. This is due to the fact that ALQASIM focuses on answer selection and validation while the other two Arabic QA4MRE systems depend on IR redundancy based approaches instead of analyzing the reading test documents. Both IDRAAQ Abouenour et al. [6] and Trigui et al. [7] depend on passage retrieval and do not give enough attention to answer selection and validation, which is not efficient enough in handling QA4MRE.

In our future research, we will focus on applying rule-based techniques that depend on Arabic specific syntactic and semantic rules, which proved to be effective in other NLP tasks like Arabic sentiment analysis [14]. Testing our work with multilingual QA, especially cross-language QA, will also be of great benefit, so as to make use of the challenges solved in that field [15]. We will also work on using feature selection with text classification as explained by Saleh and El-Sonbaty [16], to build specialized ontologies automatically from the content found on the Internet, to help with the semantic expansion and disambiguation of the different answer choices.

Bibliography

[1] Peñas, A., Hovy, E., Forner, P., Rodrigo, A., Sutcliffe, R., Sporleder, C., Forascu, C., Benajiba, Y., & Osenova, P. (2012, September). Overview of QA4MRE at CLEF 2012: Question Answering for Machine Reading Evaluation. In *CLEF 2012 Workshop on Question Answering For Machine Reading Evaluation (QA4MRE)*.

[2] Peñas, A., Hovy, E. H., Forner, P., Rodrigo, Á., Sutcliffe, R. F., Forascu, C., & Sporleder, C. (2011). Overview of QA4MRE at CLEF 2011: Question Answering for Machine Reading Evaluation. In *CLEF (Notebook Papers/Labs/Workshop)*.

[3] Penas, A., & Rodrigo, A. (2011, June). A simple measure to assess non-response. In *Proceedings of the 49th Annual Meeting of the Association for Computational Linguistics: Human Language Technologies*-Volume 1 (pp. 1415–1424). Association for Computational Linguistics.

[4] Ko, J., Si, L., & Nyberg, E. (2007). A Probabilistic Framework for Answer Selection in Question Answering. In *HLT-NAACL* (pp. 524–531).

[5] Mendes, A. C., & Coheur, L. (2011, June). An Approach to Answer Selection in Question-Answering Based on Semantic Relations. In *IJCAI* (pp. 1852–1857).

[6] Abouenour, L., Bouzoubaa, K., & Rosso, P. (2012, September). IDRAAQ: New Arabic Question Answering System Based on Query Expansion and Passage Retrieval. In *CLEF 2012 Workshop on Question Answering For Machine Reading Evaluation (QA4MRE)*.

[7] Trigui, O., Belguith, L. H., Rosso, P., Amor, H. B., & Gafsaoui, B. (2012). Arabic QA4MRE at CLEF 2012: Arabic Question Answering for Machine Reading Evaluation. In *CLEF (Online Working Notes/Labs/Workshop)*.

[8] Bhaskar, P., Pakray, P., Banerjee, S., Banerjee, S., Bandyopadhyay, S., & Gelbukh, A. (2012, September). Question Answering System for QA4MRE@CLEF 2012. In *CLEF 2012 Workshop on Question Answering For Machine Reading Evaluation (QA4MRE)*.

[9] Ezzeldin, A. M., Kholief, M. H., & El-Sonbaty, Y. (2013). ALQASIM: Arabic language question answer selection in machines. In *Information Access Evaluation. Multilinguality, Multimodality, and Visualization* (pp. 100–103). Springer Berlin Heidelberg.

[10] Habash, N., Rambow, O., & Roth, R. (2009, April). Mada+ tokan: A toolkit for arabic tokenization, diacritization, morphological disambiguation, pos tagging, stemming and lemmatization. In *Proceedings of the 2nd International Conference on Arabic Language Resources and Tools (MEDAR)*, Cairo, Egypt (pp. 102–109).

[11] Oraby, S., El-Sonbaty, Y., & Abou El-Nasr, M. (2013, October). Exploring the Effects of Word Roots for Arabic Sentiment Analysis. In *6th International Joint Conference on Natural Language Processing*, Nagoya, Japan, October 14–18, 2013.

[12] Khoja, S., & Garside, R. (1999), *Stemming Arabic Text*, Computing Department, Lancaster University, Lancaster, UK.

[13] Taghva, K., Elkhoury, R., & Coombs, J. (2005, April). Arabic Stemming without a Root Dictionary. In *International Conference on Information Technology: Coding and Computing*, 2005. ITCC 2005. (Vol. 1, pp. 152–157). IEEE.

[14] Oraby, S., El-Sonbaty, Y., & El-Nasr, M. A. (2013). Finding Opinion Strength Using Rule-Based Parsing for Arabic Sentiment Analysis. In *Advances in Soft Computing and Its Applications* (pp. 509–520). Springer Berlin Heidelberg.

[15] Forner, P., Giampiccolo, D., Magnini, B., Peñas, A., Rodrigo, Á., & Sutcliffe, R. (2010). Evaluating multilingual question answering systems at CLEF. *Target*, 2(008), 009.

[16] Saleh, S. N., & El-Sonbaty, Y. (2007, November). A feature selection algorithm with redundancy reduction for text classification. In *22nd international symposium on Computer and information sciences*, 2007. iscis 2007. (pp. 1–6). IEEE.

Andrea Bellandi, Alessia Bellusci, and Emiliano Giovannetti

Computer Assisted Translation of Ancient Texts: The Babylonian Talmud Case Study

Abstract: In this paper, we introduce some of the features characterizing the Computer Assisted Translation web application developed to support the translation of the Babylonian Talmud (BT) in Italian. The BT is a late antique Jewish anthological corpus, which, as other ancient texts, presents a number of hurdles related to its intrinsic linguistic and philological nature. In this work, we illustrate the solutions we adopted in the system, with particular emphasis on the Translation Memory and the translation suggestion component.

1 Introduction

Computer-Assisted Translation (CAT) tools are conceived to aid in the translation of a text. At the very heart of a CAT tool there is always a Translation Memory (TM), a repository that allows translators to consult and reuse past translations, primarily developed to speed up the translation process. As the use of TMs spread during the 1990s, the market quickly shifted from the reduction of translation time to the reduction of costs, Gordon (1996). For example, phrases having an exact correspondence inside the TM were charged at lower rates. However, considering the nature of the texts we are working on (mainly ancient texts of historical, religious and cultural relevance), It is mandatory to safeguard the quality of the translation, beside optimizing its speed.

To translate an ancient text there is the need of (at least) two kinds of competences, both on the languages (source and target) and the "content". In this perspective users shall wear at the same time the hat of the translator and the scholar. Hence, a system developed to support the translation of ancient texts shall go beyond the standard set of functionalities offered by a CAT tool.

In this paper we discuss the experience matured in the context of the "Progetto Traduzione del Talmud Babilonese", monitored by the Italian Presidency of the Council of Ministers and coordinated by the Union of Italian Jewish Com-

Andrea Bellandi, Alessia Bellusci, Emiliano Giovannetti: Istituto di Linguistica Computazionale, Consiglio Nazionale delle Ricerche, via G. Moruzzi 1, 56124 – Pisa, Italy, e-mail: {andrea.bellandi, alessia.bellusci, emiliano.giovannetti}@ilc.cnr.it

munities and the Italian Rabbinical College. In the context of the Project we are in charge of developing a collaborative web-based application to support the translation of the Babylonian Talmud (BT) into Italian by a group of authorized translators.

It is not the purpose of this paper to describe the full CAT system we are working on, which includes, among other things, semantic annotation features. Instead, we will focus here on the development of the TM and on the suggestion component, both designed for a community of translators working, in a collaborative way, on ancient texts – such as the BT –, whose plain and literal translation shall be enhanced to be intelligible. In this paper we will emphasize the features of our system w.r.t. those of the available open source CAT tools (*OpenTM*, *OmegaT*, *Transolution*, *Olanto*).

In Section 2 we discuss the peculiarities of the BT and how they fit for the BT translation by means of a TM. Section 3 describes the construction process of the TM, and presents all the features of our CAT system. Section 4 provides a preliminary evaluation of its performance. Finally, Section 5 outlines conclusions and future works.

2 The Babylonian Talmud

In the next sections, we provide a brief historical and linguistic analysis of the BT and discuss the relevant features of this rich textual corpus in relation to the development of a TM aimed at supporting its translation in modern languages.

2.1 Historical and Content Analysis

The Babylonian Talmud (BT) represents the most important legal source for Orthodox Judaism and the foundation for all the successive developments of *Halakhah* and *Aggadah*. Compiled in the Talmudic academies of late antique Babylonia (contemporary Iraq), the BT is based on sources from different epochs and geographical areas and maintained a fluid form at least until the sixth century CE. The BT corresponds to the effort of late antique scholars ("*Amoraim*") to provide an exegesis of the *Mishnah*, an earlier rabbinic legal compilation, divided in six "orders" ("*sedarim*") corresponding to different categories of Jewish law, with a total of 63 tractates ("*massekhtaot*"). Although following the inner structure of the *Mishnah*, the BT discusses only 37 tractates, with a total of 2711 double sided folia in the printed edition (Vilna, XIX century). The BT deals with ethics, jurisprudence, liturgy, ritual, philosophy, trade, medicine, astronomy, magic and so much more.

To this conspicuous volume of textual material and topics corresponds an intricate complex of heterogeneous sources. The BT includes (i.) quotations from the *Mishnah* ["*Mishnah*"], (ii.) long *amoraic* discussions of *mishnaic* passages aimed at clarifying the positions and lexicon adopted by the *Tannaim* ["*Gemarah*"], and (iii.) external *tannaitic* material not incorporated in the canonical *Mishnah* ["*Baraytot*"]. Intrinsically related to the *Mishnah*, the *Gemarah* develops in the form of a dialectical exchange between numerous rabbinic authorities belonging to different generations, epochs (*Tannaim* and *Amora'im*), and geographical areas (Graeco-Roman Palestine and Sassanid Babylonia). The thorough exegetical work on the *Mishnah* operated in the *Gemarah* implies discussing the words and phrases adopted in the *Mishnah*, speculating on the principles behind the *Mishnah* case laws, linking the different rules stated in the *tannaitic* text to the Bible, and harmonizing the contradictions occurring in the *Mishnah*. The *sugya'* represents the basic literary unit of the *Gemarah*, which displays the relevant textual material in a succession of questions and answers. To each *mishnah* ["statement"/ "law" from the *Mishnah*] corresponds one or more *sugyot*. Although constructed according to specific literary conventions, the succession of *sugyot* in the BT gives the perception of a fluent and live debate between rabbinic authorities. The BT attests to an extremely vast body of legal ("*halakhic*") and narrative ("*aggadic*") knowledge, which had been continuously transmitted and re-interpreted in late antique Judaism.

2.2 Linguistic and Stylistic Analysis

To the content and philological complexity of the BT, we ought to add the linguistic richness presented by this textual corpus, which inevitably affected our choices for the development of the web application for supporting its translation. In its extant form, in fact, the BT attests to (i.) different linguistic stages of Hebrew (Biblical Hebrew, Mishnaic Hebrew, Amoraic Hebrew), (ii.) different variants of Jewish Aramaic (Babylonian Aramaic and Palestinian Aramaic), and (iii.) several loanwords from Akkadian, ancient Greek, Latin, Pahlavi, Syriac and Arabic. In addition, all the different languages and dialectal variants are often alternated in the text, thus making difficult, if not impossible, to develop an automatic (statistical or pattern-based) sentence splitter (see Segmentation Process subsection).

To date, there are no available Natural Language Processing (NLP) tools suitable for processing ancient North-western Semitic languages, such as the different Aramaic idioms attested to in the BT, and for detecting the historical variants of the Hebrew language as used in the Talmudic text. The only existing NLP tools for Jewish languages (see Bar-Haim et al. (2005), Itai (2006), *HebMorph*[1]) are specifically implemented for Modern Hebrew, a language which has been artificially revitalized from the end of the XIX century and which does not correspond to the idioms recurring in the BT. In its multifaceted form the "language" of the BT is unique and attested to only in few other writings. In addition, only few scholars have a full knowledge of the linguistic peculiarities of the BT and even fewer experts in Talmudic Studies are interested in collaborating to the creation of computational technologies for this textual corpus. Therefore, the development of NLP tools for BT would require a huge and very difficult effort, which would probably not be justified by the subsequent use of the new technologies developed. Also for these reasons, we opted to a translation system based on a TM.

More important, the literary style of the BT itself concurred to orientate us through the choice of the TM. Composed in a dialogical form, the BT is characterized by formulaic language and by the recurrence throughout the different tractates of fixed expressions used to connect the different textual units. As an exegetical text, the BT contains innumerable quotations from the *Mishnah*, from other *tannaitic* sources and even from *amoraitic* statements discussed in other passages of the BT itself. In addition, since one of the major efforts of the *Amora'im* was finding the relevant Biblical passages on which to anchor the rules of the *Mishnah*, the BT turns out to be a constellation of quotations from the Scripture (see Quotation subsection). The exceptional volume of citations and recurrent fixed expressions in the BT makes the TM particularly effective and fit for this writing.

2.3 Statistics and Comparison with Other Texts

In the course of our study we carried out also some statistic analyses on the language and style of the BT, which corroborated the choice of the TM approach. Particularly, the evaluation of the lexical variation within five of the BT tractates (*Berakhot*, *Shabbat*, *Rosh Ha-Shanah*, *Qiddushin*, and *Bava Metziah*) suggested that its lexicon is relatively poor when compared to other Jewish texts (see, Table 1).

The Type-Token Ratio (TTR), expressing the lexical variety, was estimated for every tractate and was computed as the ratio between the number of the distinct word tokens (ignoring the repetitions) and the number of all tokens in the analyzed text. A high TTR indicates, in fact, a high degree of lexical variations, while

1 Morphological Analyser and Disambiguator for Hebrew Language http://code972.com/hebmorph

Tab. 1: Text statistics.

Text	Tokens	Types	Average TTR	RR
Rosh Hashana	23,660	5,758	0.72	0.13
Berakhot	70,622	13,355	0.70	0.10
Qiddushin	59,586	10,171	0.70	0.11
Shabbat	114,082	19,096	0.70	0.11
Bava Metzia	84,675	12,774	0.70	0.10
Liqra't Medinah	4,772	2,444	0.84	0.39
Bi-Vedidut	11,552	7,230	0.90	0.46
Arieh Baal Guf	12,653	6,417	0.89	0.45
Sefer Ha-Kuzari	49,617	20,680	0.87	0.30
Sefer Ha-Bahir	15,759	5,546	0.79	0.24
Sefer Ha-Razim	7,067	3,545	0.87	0.37

a low TTR demonstrates relatively low lexical variation. In our comparison, we computed TTR regardless of the text length, by taking a moving average type token ratio (MATTR) Covington and McFall (2010). By using a 100-words sized window, we computed TTRs for each window in the text and averaged them. According to the results reported in the column "Average TTR" in Table 1 (grey part), all the tractates present a poor vocabulary, containing highly repetitive words/phrases. Furthermore, the ratio between the number of hapax legomena and the number of all tokens (RR) is lower in the BT tractates, suggesting a high degree of lexical repetition.

2.4 Relevant Aspects of the BT for the CAT System

The complexity and richness of the BT makes this textual collection an exceptional object of study for many aspects (textual content; different languages attested to; history of the composition of the text; sources employed; passage from oral transmission to textual redaction; philology of the text), and, yet highly increases the hardship of analysing and translating it.

An ancient anthological writing such as the BT cannot be treated and translated as a modern text, since a plain and literal translation of this text would not be intelligible to a modern reader. The Talmudic text is exceptionally concise and several passages remain unclear even for expert Talmudists. Therefore, a good translation of the BT requires the addition of explicative integrations within the translation and a re-elaboration of the literary tradition. For these reasons we had to develop a system suitable for translators, who have a solid background in Talmudic Studies, and to create additional tools capable of expressing all the features of a translation, which is not merely a translation, but, to a certain extent, an interpretative commentary itself. Particularly, we had to enhance the system with tools that enable to distinguish the literal part of the translation (indicated in bold) from the explicative additions of the translators/scholars (see 3.1). Furthermore, due to the complexity of the inner structure of the BT, we had to provide translators with the possibility to organize their translations in specific units and subunits (see Structure subsection).

In addition, the peculiar literary style of the BT (formulaic language, standard expressions, quotations from the Bible and *tannaitic* sources, quotations from other sections of the BT itself) together with the lack of NLP tools for the languages attested to in the BT, oriented us, in the first place, to base our translation system on a TM instead of Machine Translation approaches.

3 Features of the CAT System

In the following sections we outline the most relevant features developed to face the translation of textual corpora with complex philological and linguistic peculiarities, such as the BT.

3.1 The Translation Memory

Our TM is organized at the segment level. A segment is a portion of original text having an arbitrary length. We formally define the TM $M_{BT} = \{s_i, T_i, A_i, c_i\}$ with i ranging from 1 to n, as a set of n tuples, where each tuple is defined by:

- s_i, the source segment;
- $T_i = \{t_i^1, \ldots, t_i^k\}$, the set of translations of s_i with $k \geq 1$, where each t_i^j includes a literal part $\vec{t}_i^j$ exactly corresponding to the source segment, and an explicative addition, from now on referred to as contextual information $\overleftarrow{t}_i^j$, with $1 \leq j \leq k$;
- $A_i = \{a_i^1, \ldots, a_i^k\}$, the set of translators' identifiers of each translation of s_i in T_i with $k \geq 1$;
- c_i, the context of s_i referring to the tractate to which s_i belongs.

As stated above, the translation of each segment can be done by differentiating the "literal" translation (using the bold style) from explicative additions, i.e. "contextual information". Segments having the same literal part can then differ by their contextual information. The tractate of the source segment of each translation is called "context".

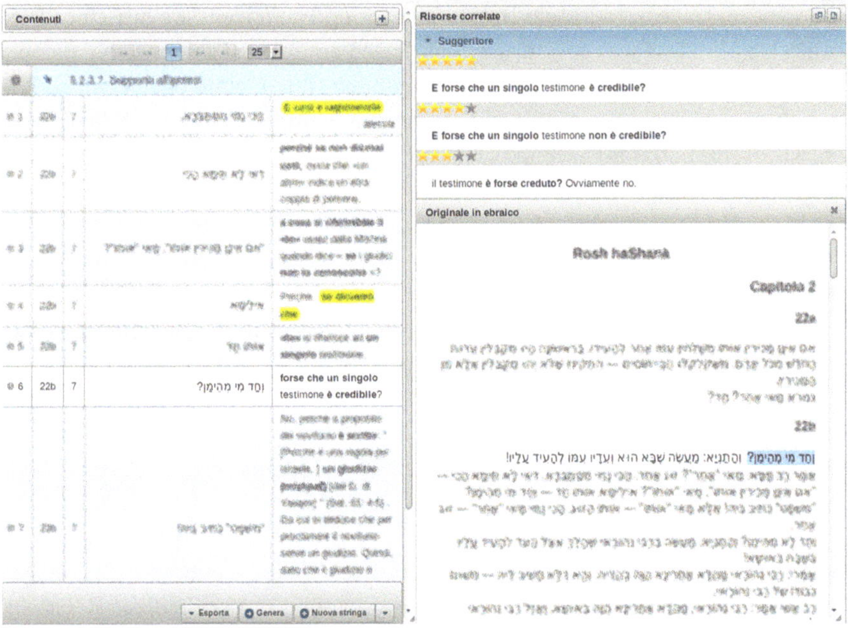

Fig. 1: The system GUI (data, authors, parts of the original Babylonian Talmud text have been clouded for privacy and rights reasons).

The Translation Suggestion Component. The system acquires the segment to be translated, queries the TM, and suggests the Italian translations related to the most similar strings. The translation procedure involves three main aspects:

- how the system allows translators to choose a source segment (see Segmentation Process subsection);
- how the system retrieves pertinent translation suggestions (see Similarity Function subsection);
- how the system presents the suggestions to each translator (see Suggestions Presentation subsection).

In the following subsections we will analyze these aspects, emphasizing all the characteristics that we consider peculiar for ancient texts and which, to our knowledge, differentiate our system from state of the art open source CAT tools.

Similarity Function. For the reasons stated in Section 2.2, it has not been possible, so far, to include either grammatical or syntactic information in the similarity search algorithm. So, we took account of adopting similarity measures based on edit distance, $ED(s_1, s_2)$, by considering two segments to be more similar when the same terms tend to appear in the same order. The novelty we introduce here consists in the way we rank suggestions, based on external information, namely i) the authors of translations and ii) the context (the tractate of reference). The information about i) the author of the translation and of ii) the tractate of reference is useful both for translators and revisors. On the one hand, translators are enabled to evaluate the reliability of the suggested translations on the basis of the authority and expertise of the relative translators. On the other hand, revisors can exploit both kinds of information to ensure a more homogeneous and fluent translation.

Given a segment s_q of length $|s_q|$, and a distance error δ, our similarity function allows to:

- retrieve all segments s in the TM (called suggestions) such that $ED(s_q, s) \leq round(\delta * |s_q|)$;
- rank suggestions, not only on the basis of the *ED* outcome, but also on both the current context and the suggestion author.

Searching for segments within δ errors, where δ is independent of the length of the query, could be little meaningful, as described in Mandreoli et al. (2002). For this reason, we considered δ as the percentage of admitted errors w.r.t. the sentences to be translated, multiplying it by the length of the query segment. In collaboration with the translators' team, we have experimentally tuned δ to 0.7. Our algorithm is based on dynamic programming, and its implementation refers to Navarro (2001). In order to compute $ED(s_1, s_2)$, it builds a matrix $M_{(0..|s_1|,0..|s_2|)}$, where each element $m_{i,j}$ represents the minimum number of token mutations required to transform $s_1(1..i)$ in $s_2(1..j)$. The computation process is the following:

$$m_{i,j} = \begin{cases} i & \text{if } j = 0 \\ j & \text{if } i = 0 \\ m_{(i-1,j-1)} & \text{if } s_1(i) = s_2(j) \\ 1 + \mu & \text{if } s_1(i) \neq s_2(j) \end{cases}$$

where $\mu = min(m_{(i-1,j)}, m_{(i,j-1)}, m_{(i-1,j-1)})$, and the final cost is represented by $m_{(|s_1|,|s_2|)}$. The system returns those strings translated in Italian that have the lower costs. Basically, given a segment to translate, many other source segments can be exactly equal to it and each of them can be paired with multiple translations in the TM. Differently from what happens with other CAT, our similarity function immediately provides the translator with more accurate suggestions, by ranking all of them on the basis of the current context, showing first those already provided by the most authoritative translators. This aspect positively affects also the revision activity. It possible, for example, to easily retrieve translations of equal segments having the same context provided by different translators in order to analyze and, in case, uniform stylistic differences. It is noteworthy that the translations' revision activity, performed by human reviewers, guarantees that the TM contains only high-quality translations.

Suggestions Presentation. The system user interface shows each suggestion marked by a number of stars, as shown in Figure 1. That number is assigned by the system on the basis of how fuzzy the match between source segments is: five stars are regarded as a perfect suggestion (exact match); four stars indicate few corrections required to improve the suggestion (fuzzy match); three stars refer, in most of the cases, to acceptable suggestions (weak fuzzy match). As stated in the previous subsection (Similarity Function), while dealing with suggestions ranked with the same number of stars, the system orders them by context and author.

Every suggestion can be shown with or without its contextual information; literal translation is specified by bold font, w.r.t. the contextual information. Each translator, for example, can then approve as correct the literal translation $\bar{t}_i^j$, and modify only $\tilde{t}_i^j$. Then, each new correct translation is added to the TM, thus, increasing the pool of available translations. In this way, our system leaves human translators in control of the actual translation process, relieving them from routine work. Our system allows also to maintain the translation process as creative as necessary, especially in those cases requiring human linguistic resourcefulness.

Figure 1 shows a simple example of translation suggestions[2]. The related set of suggestions are provided and scored[3] by the system, as depicted in Figure 1:

- t_1 = "**And may a signle** witness **be credible ?**", with $\tilde{t}_1$ = {witness}
- t_2 = "**And may a single** witness **not being credible ?**", with $\tilde{t}_1$ = {witness}
- t_3 = "**Is the** witness **believed ?** Obviously not", with $\tilde{t}_1$ = {witness, obviously not}

In the translation session of Figure 1, the translator chose t_1, and changed its literal part by deleting the word "**And**"[4].

In our view, these requirements are much more fundamental for the translation of ancient texts than for modern texts. To the best of our knowledge, there are no state of the art CAT tools taking into account these needs.

2 Clearly, the suggestions are in Italian. We are attempting to provide English translations as semantically similar to the original Italian sentences as possible.

3 For privacy reasons we omitted both context and authors.

4 Note that $\bar{t}_1$, $\bar{t}_2$, $\bar{t}_3$ are not grammatically correct in English, while they are perfectly well-formed in Italian, since Italian is a null-subject language.

3.2 Additional Tools

Segmentation Process. In all CAT systems, the first step of the translation process is the segmentation of the source text. As stated in 2.2, the BT does not exhibit a linguistic continuity, thus preventing an automatic splitting into sentences. To face this uncommon situation we opted for a manual segmentation taking into account also the structure of the text (see Structure subsection). Each translator selects the segment to translate from the original text contained in a specific window (Figure 1). This process reveals to have a positive outcome: translators, being forced to manually detect the segments and organize them in a hierarchical structure, acquire a deeper awareness of the text they are about to translate. Clearly, the manual segmentation implies the engagement of the translators in a deep cognitive process aimed to establish the exact borders of a segment. The intellectual process involved in the textual segmentation deeply affects also the final translation, by orienting the content and nature of the TM. To the best of our knowledge this feature is not present in any state of the art CAT system.

Quotations. As outlined above, the BT is based on various sources and contains quotations of different Jewish texts (Bible, *Mishnah*, *Toseftah*, Babylonian Talmud, Palestinian Talmud). For this reason, the system provides precompiled lists containing all these texts, so that users can select the relevant source related to their translation.

Structure. Since the translation has to be carried out in respect of the inner structure of the BT, we had to implement a component to organize the translation following a hierarchic subdivision: tractates, chapters, blocks (sugyot), logical units (e.g. thesis, hypothesis, objections, questions, etc.), and strings, the latter representing the translation segments.

4 Evaluation

With our system not only we aim to increase the translation speed, but also – and especially – to support the translation process, offering the translators a collaborative environment, in which they can translate similar segments without losing contextual information on the translations. Under these premises it is quite hard to exactly evaluate the overall performance of the proposed system. In order to accomplish this task, we tried to measure both the number of similar segments

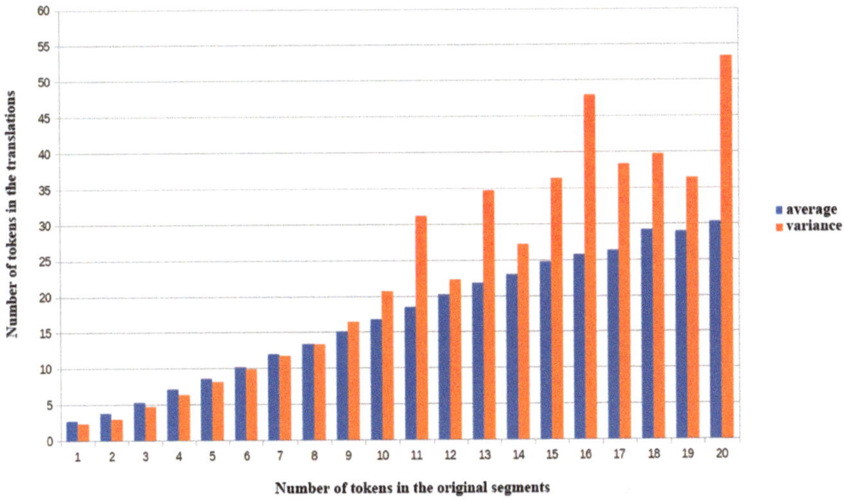

Fig. 2: Ratio between the length of the original segments and the related translations' length

stored within the TM, (i.e., the rate of TM redundancy), and how good the suggestions provided by the system are.

4.1 Translation Memory Analysis

For privacy reasons, we cannot publish the number of translated sentence present, at the moment, inside the TM. Let us say we have n translated sentences, with an average length of five tokens. Figure 2 shows that each translation is approximately between 70% and 40% longer than its original text segment (these percentages are computed as the ratio between each s_i length and the average length of all translations belonging to the related T_i). This could be partially ascribed to the fact that semitic languages are agglutinated, implying that every token is made up by the union of more morphemes. Also, the high degree of variance should be taken into account; starting from eight tokens long source segments, the variance is always higher than the average, and sudden peaks can be observed examining source segments longer than eleven tokens. This means that a significant part of the obtained translations are exclusively literal, while another significant part is enriched with contextual information, derived from translators' knowledge and scholarly background, as well as from their personal interpretations of the text.

To study the redundancy, we conducted a jackknife experiment to roughly estimate the TM performance, as in Wu (1986). We partitioned the TM in groups

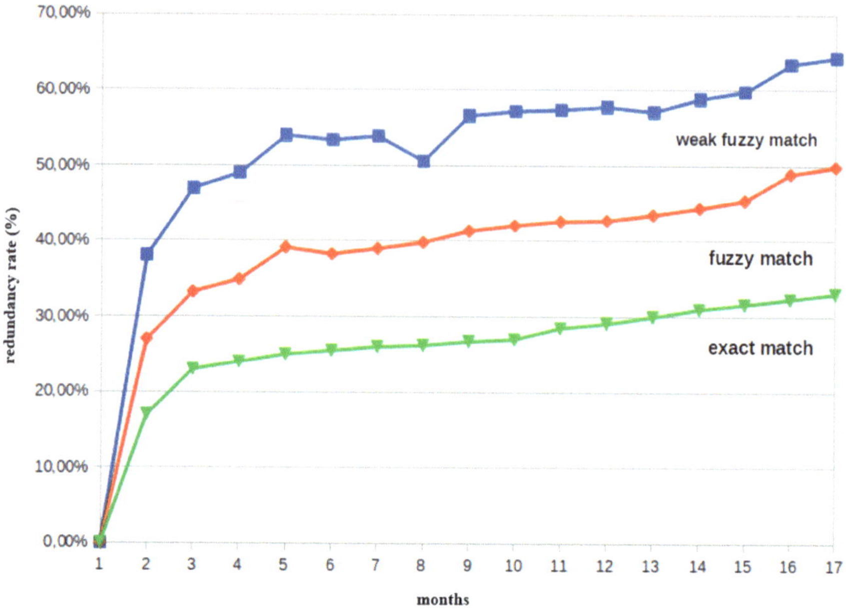

Fig. 3: TM performance.

of one thousand, then, used one set as input and translated the sentences in this set using the rest of the TM. This was repeated n times. Intuitively, the resampling method we adopted uses the variability in the test sets to draw conclusions about statistical significance. Figure 3 shows the redundancy rate of the TM from the beginning of the project until today. Redundancy curves are drawn by considering the ranking of the similarity function. The percentage of source segments found in the TM, grows logarithmically with time (and consequently with the size of the memory). Clearly the larger the memory will be, the better the performance. For instance, if we suppose that exact matches provide at least one perfect suggestion, then each translator would save 30% of their work.

4.2 Translation Suggestions Goodness

In order to measure the goodness of a translation suggestion, we used machine translation evaluation metrics. These metrics measure the similarity between a reference translation and a set of candidate ones; clearly, the more each candidate resembles the reference, the better the score. Our test has been built by identifying 1000 segments from the TM such that every correspondent T_i contains at least

Tab. 2: Translation suggestions evaluation by means of MT metrics. For each T_i, *max* represents the average of the best matches, and *min* means the average of the fuzzier ones.

MT measure		*with* contextual information	*without* contextual information
WER	*max*	0.78	0.76
	min	0.32	0.27
BLEU	*max*	0.81	0.78
	min	0.13	0.15

two identical literal translation from different authors ($\bar{t}_i^h = \bar{t}_i^k$ with $h \neq k$, and $a_i^h \neq a_i^k$). We considered each of those translations as the reference translation t_i^{ref} of the related source segment s_i^{test}, with i ranging from 1 to 1000. We removed $\{s_i^{test}\}$ set from the TM, considering the remaining set of translations as the candidates set. Given the large number of existing evaluation metrics, we focused on those we retain the two more representative ones in the literature: Word Error Rate (WER), and BLEU, described in Papineni et al. (2002). WER, one of the first automatic evaluation metric used in machine translation, is computed at word level by using the Levenshtein distance between the candidate and the reference, divided by the number of words in the reference[5]. BLEU, instead, measures n-gram precision, i.e., the proportion of word n-grams of the candidate that are also found in the reference. We produced our own implementations of WER and BLEU, by normalizing WER values between 0 and 1. For each s_i^{test}, our similarity function retrieved the best match (if available), and each of its related Italian translations has been compared with t_i^{ref}, saving both the best and worst scores for each measure. Finally, all the best scores and the worst ones, have been averaged and reported in Table 2 as *max* and *min* values, respectively. This process has been performed on both literal translations and translations with contextual information. Concerning *max* values, regardless the presence of contextual information, the BLEU score is greater than WER. WER counts word errors at the surface level. It does not consider the contextual and syntactic roles of a word. This result is encouraging since BLEU, instead, gives us a measure of the translations' quality, as reported in White et al. (1993) considering two different aspects: lower-order n-grams tend to account for adequacy, while higher-order n-grams tend to account for fluency (we used n-grams with n ranging from 1 to 4).

5 Several variants exist, most notably TER (Translation Edit or Error Rate), in which local swaps of sequences of words are allowed. However, WER and TER are known to behave very similarly, as described in Cer et al. (2010).

Instead, when we examine *min* values, the results are completely reversed. Both measures indicate that values are very low: in particular BLEU shows that the quality of the provided translations is very poor (15%). This situation brought us to repeat the BLEU measurement taking into account only the exact matches of the test set: in this case, the minimal values arise to the 40%, still showing a bad quality of some translations. Based on these results, we believed necessary to integrate a revision activity for the translators' work, as it was originally scheduled within the project.

Another information we can deduce from Table 2 is that, for both measures, the translations accounting for contextual information achieve better results than those achieved by literal ones. The reason for this phenomenon could depend on the fact that, as stated in Firth (1957), similar words or similar linguistic structures, tend to occur in similar contexts. Therefore, we can assume that similar literal translation parts present similar contextual information parts. Indeed, in the previous example, we can see that the word "witness" is a contextual information that appears in all the provided suggestions (see "testimone" in Figure 1).

5 Conclusions and Future Works

In this paper, we introduced some components of the CAT system we developed to support the collaborative translation of the Babylonian Talmud in Italian. Differently from the functionalities a CAT system typically offers to ease and speed up the translation of a text, the translation (and revision) of an ancient text as the BT requires specific features. In general, the environment must create the conditions to allow a user, in the double role of translator and scholar, to produce correct translations, which often require, especially for texts such as the BT, several textual additions. We opted for a TM instead of a Machine Translation approach for a number of reasons, both related to the intrinsic nature of the BT (high number of quotations, recurrence of fixed expressions and dialectical narration) and to the lack of NLP tools for ancient Semitic languages. We focused here on the description of the Translation Memory and on some aspects regarding the relative translation suggestion component and its evaluation. Among the most interesting features we introduced in the system, there are: i) the manual segmentation, necessary to compensate to the lack of punctuation marks, required for automatic sentence splitting, ii) translation suggestions, which take into account both the author of the translation and the context (very important also for the revision process), iii) the possibility to distinguish between literal translations and explicative additions, iv) the possibility to add references to quotations, v) the organization of the translations into hierarchical subdivisions. Concerning future works, we plan to automatically populate the TM with biblical quotations, taking into account the work described in HaCohen-Kerner (2010), and to integrate a Language Identification tool both to improve the performance of the suggestion component and to allow advanced searches on the translated corpus on a language basis. Furthermore, we are working on the development of a language independent, open source and web-based CAT system, since we believe most of the solutions adopted for the translation of the BT could be generalized to assist in the translation of the majority of ancient texts.

Acknowledgements

This work has been conducted in the context of the research project TALMUD and the scientific partnership between S.c.a r.l. “Progetto Traduzione del Talmud Babilonese” and ILC-CNR and on the basis of the regulations stated in the “Protocollo d’Intesa” (memorandum of understanding) between the Italian Presidency of the Council of Ministers, the Italian Ministry of Education, Universities and Research, the Union of Italian Jewish Communities, the Italian Rabbinical College and the Italian National Research Council (21/01/2011).

Bibliography

Bar-Haim, R., Sima’an, K. and Winter, Y. (2005) Choosing an Optimal Architecture for Segmentation and POS-Tagging of Modern Hebrew. In: *Proceedings of the Association for Computational Linguistics Workshop on Computational Approaches to Semitic Languages*, 2005, pp. 39–46.

Cer, D., Manning, C. D., and Jurafsky, D. (2010) The Best Lexical Metric for Phrase-Based Statistical MT System Optimization. In: *Human Language Technologies, the* 11th *Conference of the North American Chapter of the Association for Computational Linguistics*, Los Angeles, USA, June 2–4, 2010, pp. 555–563.

Covington, M.A., and McFall, J.D. (2010) Cutting the Gordian Knot: The Moving-Average Type-Token Ratio (MATTR). *Journal of Quantitative Linguistics*, 17(2):94–100.

Firth, J. R. (ed.), *Papers in Linguistics 1934*. Oxford University Press, London.

Gordon, I. (1996) Letting the CAT out of the bag - or was it MT?. In: *Proceedings of the* 8th *International Conference on Translating and the Computer*, Aslib, London, 1996.

HaCohen-Kerner, Y., Schweitzer, N., and Shoham, Y. (2010) Automatic Identification of Biblical Quotations in Hebrew-Aramaic Documents. In: *Proceedings of the International Conference on Knowledge Discovery and Information Retrieval*, Valencia, Spain, October 25–28, pp. 320–325.

Itai, A. (2006) Knowledge Center for Processing Hebrew In: *Proceedings of the 5th International Conference on Language Resources and Evaluation – Workshop "Towards a Research Infrastructure for Language Resources"*, Genoa, Italy, 22 May, 2005.

Mandreoli, F., Federica M., Martoglia, R., and Tiberio, P. (2002) Searching Similar (Sub)Sentences for Example-Based Machine Translation. In: *Italian Symposium on Advanced Database Systems (SEBD)*, Portoferraio (Isola d'Elba), Italy, 19–21 June, 2002.

Navarro, G. (2001) *A guided Tour to Approximate sentence Matching*, Technical report, Department of Computer Science, University of Chile.

Papineni, K., Roukos, S., Ward, T., and Zhu, W.J. BLEU: A Method for Automatic Evaluation of Machine Translation. In: *Proceedings of the 40th Annual Meeting on Association for Computational Linguistics*, Stroudsburg, PA, USA, 2002, pp. 311–318.

White, J. S., O'Connell, T. A., and Carlson, L. M. (1993) Evaluation of Machine Translation. In: *Proceedings of the workshop on Human Language Technology Association for Computational Linguistics*, Stroudsburg, PA, USA, 1993, pp. 206–210.

Wu, C. F. J. (1986) Jackknife, Bootstrap and Other Resampling Methods in Regression Analysis. *The Annals of Statistics* 14(4):1261–1295.

Jesús Calvillo and Matthew Crocker

A Rational Statistical Parser

Abstract: A model of syntactic parsing that combines elements of information and probability theory is proposed. The model assigns probability and entropy scores to parse trees: trees with higher probabilities are preferred while trees with higher entropies are penalized. This model is argued to be psycholinguistically motivated by means of *rational analysis*. Using a grammar extracted from the Penn Treebank, the implemented model was evaluated on the section 23 of the corpus. The results present a modest but general improvement in almost all types of phenomena analyzed, suggesting that the inclusion of entropy is beneficial during parsing and that, given our formulation, its relevance decreases polynomially as the syntactic trees develop during incremental processing.

1 Introduction

John Anderson (1991) proposes a framework to study cognitive systems dubbed as *rational analysis*. The main assumption is that humans are adapted to their environment and goals, and therefore if there is an optimal path to the solution of a task, the human cognitive system should converge towards a similar solution. This means that for some tasks, if we want to model human behavior, it is useful to first identify an optimal solution given resource limitations and the environment; this solution should be similar to the one that humans would apply.

Applying this rationale to syntactic parsing, we propose a framework for parsing that integrates probability estimations of parse derivations with predictions about unseen parsing events in the form of entropy. Models within this framework will be argued to be accomplishing three rational goals: a)retrieving the most probable analyses according to experience, b) being quick in order to cope with communication time constraints, and c)retrieving analyses minimizing cognitive costs.

The organization will be as follows: section 2 corresponds to some background information, section 3 presents the main ideas of our parsing model,

Jesús Calvillo and Matthew Crocker: Deparment of Computational Linguistics and Phonetics, Saarland University, Germany, e-mail: jesusc@coli.uni-saarland.de, crocker@coli.uni-saarland.de

section 4 corresponds to the experiments, and we end with the conclusion in section 5.

2 Background

Chater et al. (1998) apply the rational analysis framework to formulate a model of human syntactic processing. According to them, one goal of the parser is to retrieve the correct parse minimizing the probability of either failing during extensive backtracking or mistakenly dismissing the correct parse. Thus, the human syntactic parser follows the best parsing paths according to a parsing score $f(\phi)$ consisting of: the probability of the path $P(\phi)$, the probability of settling on that path $P(settle\ \phi)$ and the probability of escaping from that path if it is incorrect $P(escape\ \phi)$. These are traded off to achieve a sequence of how the parser should explore the parsing space maximizing the probability of success.

$$f(\phi) = P(\phi) \cdot P(settle\ \phi) \cdot \frac{1}{1 - P(escape\ \phi)} \quad (1)$$

Hale (2011) also applies rational analysis to parsing, obtaining a formulation that comprises two types of cost in an A* search: the cost paid to reach the current state from a starting point $g(n)$, and the expected cost of following a specific path from the current state $\hat{h}(n)$. In this case, the goal is to retrieve the correct parse as soon as possible as an urge due to time pressures.

$$\hat{f}(n) = g(n) + \hat{h}(n) \quad (2)$$

Hale (2006) further argues that comprehenders may be sensitive to the uncertainty of the rest of the sentence, and uses the notion of Entropy of a Nonterminal as Grenander (1967) defines it, which is the same that we will use: the entropy $H(nt)$ of each non-terminal nt in a PCFG G can be calculated as the sum of two terms. Let the set of production rules in G be R and the subset rewriting the ith non-terminal nt_i be $R(nt_i)$. Denote by $P(r)$ the probability of rule r. Then:

$$H(nt_i) = - \sum_{r \in R(nt_i)} P(r) log P(r) + \sum_{r \in R(nt_i)} P(r)[H(nt_{j1}) + H(nt_{j2}) + \ldots] \quad (3)$$

The first term corresponds to the definition of entropy for a discrete random variable. The second term is the recurrence. It expresses the intuition that derivational uncertainty is propagated from children ($nt_{j1}, nt_{j2}, \ldots$) to parents (nt_i) (Hale, 2006). The solution to the recurrence can be written as a matrix equation and directly calculated from the rule probabilities of a PCFG.

Building up on these ideas, we will present a framework for parsing that will also be argued to be complying with rational analysis.

3 Parsing Framework

We can define a parsing path as a sequence of transitions between parsing states. Each state ω corresponds to the configuration that the parser has in a specific moment regarding a possible derivation of the analyzed sentence. Transitions between states are performed through actions. Each parsing state has an associated set of actions that can be possibly done from that state according to the parsing algorithm. For example, using a left-corner algorithm each parsing state corresponds to a particular state of the stack and the so far derived tree. After performing one of the operations of left-corner parsing: scan, shift or project, the parser arrives to a new state. When the parser can perform more than one action at a particular state, we say that it is at a point of ambiguity and the parser has to decide which action is preferred, which path to follow.

A parsing state ω is deemed better if it has higher probability, and/or if from that state we can predict that we will arrive to a goal state with higher probability or sooner. A goal state corresponds to a complete syntactic analysis of the sentence. These two factors can be formalized as a score $S(\omega)$:

$$S(\omega) = P(\omega) - \beta \cdot Ent(\omega) \tag{4}$$

where $P(\omega)$ is the probability of the so far derived parse tree according to a PCFG, given the past decisions; and $Ent(\omega)$ contains information, in the form of Entropy, of how the rest of the derivation is going to be from the current state to a goal state.

More precisely we will formulate our parsing score as:

$$S(\omega) = \sum_{r \in R_\omega} logP(r) - \beta \cdot \sum_{nt \in NT_\omega} H(nt) \tag{5}$$

where R_ω is the sequence of rules employed in the derivation of ω, NT_ω is the set of nonterminals contained in ω, and β is the trade off between the two addends.

3.1 Entropies, Probabilities and Lengths

In Information Theory, entropy is a measure of uncertainty in a random event (Shannon, 1948). By taking another point of view, it can also be seen as a negative expectation of a log probability. Having a nonterminal symbol nt_i of a PCFG G, there is a set of rewritings $R(nt_i)$ available for this nonterminal according to the grammar. Each of these rewritings corresponds to one grammar rule r that in turn has a probability associated $P(r)$. If we take the rewriting of the nonterminal as a discrete random variable, then the expectation of the log probabilities associated with the possible rewritings of nt_i is defined by:

$$E_i[logP(r)] = \sum_{r \in R(nt_i)} P(r)logP(r) \tag{6}$$

which is the same as first term in the previous formulation of the Entropy of a Nonterminal in (3), with the difference of the sign. This corresponds to a single rewriting of nt_i.

Applying the same reasoning to the recursion in (3) and since the recursion factors into single rewriting steps, we can draw a similar relation between the Entropy of a Nonterminal $H(nt)$ and the Expected Log Probability of a tree whose root is nt. In other words, the entropy of a nonterminal is equal to the negative expectation of the log probability of a tree whose root is the nonterminal:

$$H(nt) = -E[logP(\tau_{nt})] \tag{7}$$

Then, a parser that follows low entropy paths is not only preferring low uncertainty, but it also prefers derivations that in average have high log probabilities.

From a different perspective and according to Hale (2011), this definition of entropy of a nonterminal correlates to the estimated derivation length taken from a corpus. This is perhaps not very surprising given the way that derivation probabilities are calculated using PCFGs. In a PCFG, the probability of a tree results from the product of the rule probabilities used in the derivation. Since probabilities range from 0 to 1, the more rules applied in the derivation, the less probable the parse tree becomes. If this is true, then we can see probability as a derivation length, and entropy as expected derivation length, one related to the already seen input and the other one to predictions about future input.

Summing up, we argue that regarding parsing derivations:

- A parser that minimizes entropies is at the same time maximizing expected log probabilities and/or minimizing expected lengths.
- Log probabilities are related to lengths as expected log probabilities (negative entropies) are related to expected lengths.
- A parser that maximizes probabilities is at the same time minimizing lengths.

Thus, on the one hand our model can be seen as minimizing derivation lengths and expected derivation lengths, and therefore following the rational goal of *being quick*, just as Hale (2011). On the other hand, the model maximizes probabilities and expected probabilities and therefore follows the rational goal of *being accurate according to experience.*

3.2 Surprisal and Entropy Reduction

According to the Entropy Reduction Hypothesis (ERH), the transition from states with high entropies to states with low entropies represents a high cognitive effort as result of disambiguation work that the parser has to achieve when incorporating new information (Hale, 2006). Similarly, Surprisal is regarded as a cognitive load measure that is high when the parser is forced to process improbable events, and its related effort is argued to be the result of the disambiguation that the parser performs in view of new information (Hale, 2001).

The first addend of our model is related to the probability of the derivation so far processed. Using log probabilities, we arrive to a definition of Surprisal of a tree τ by simply changing its sign:

$$Surprisal(\tau) = -logP(\tau) \tag{8}$$

so a maximization of probabilities is at the same time a minimization of surprisal.

Noticing the form of the surprisal formulation, it is easy to see that the definition of entropy is equal to the expected value of surprisal. Hence, Roark (2011) dubbed it Expected Surprisal:

$$ExpectedSurprisal(\tau_{nt}) = -E[logP(\tau_{nt})] = H(nt) \tag{9}$$

where τ_{nt} is a tree whose root is the nonterminal *nt*.

Then we can see the similarity between surprisal and entropy reduction. Both measure how improbable an event is, or in our context, a syntactic analysis. The first one will be high when the parser follows a low probability path, and the second will be high when the parser is forced to follow paths that are expected to have low probabilities.

From this point of view, instead of maximizing probabilities, the parse score used in our model can be rewritten as a cost or cognitive load measure consisting of two elements: one related to the effort of the already processed part of the sentence (Surprisal) and another one related to possible predictions about the rest of the sentence (Expected Surprisal) during incremental processing:

$$C(\tau) = Surprisal(\tau) + ExpectedSurprisal(\tau) \tag{10}$$

Consequently, we can see our model as being on the one hand maximizing probabilities and on the other hand minimizing cognitive load. Thus, for a given sentence, the parser would assign the tree $\hat{\tau}$ the following way:

$$\hat{\tau} = \arg\max_{\tau \in T}(logP(\tau) + E[logP(\tau)]) = \arg\min_{\tau \in T}(Surprisal(\tau) + Entropy(\tau)) \tag{11}$$

where *T* is the set of all trees matching the sentence and licensed by the grammar.

If improbable analyses are more difficult to process by people, then it is rational that people should naturally follow paths with less expected cognitive load during incremental parsing. By doing so, the parser also achieves a rational goal of *retrieving the analyses using the least amount of resources*.

3.3 Implemented Model

The implemented model lies at the computational level of analysis, describing an estimation of how preferable a parse tree is, regardless of the algorithm used for its construction.

One assumption is that during production speakers are cognitively constrained and prone to generate cognitively manageable constructions. Also, in order to maximize the probability of a successful communication, the speaker should modulate the complexity of his/her utterances in a way such that the comprehender is able to follow, as Jaeger and Levy (2006) suggest. If this is true, we can expect that normal/understandable sentences should show a bias towards cognitively manageable constructions, avoiding peaks in cognitive load.

In this model, each parsing state corresponds to a parse (sub)tree τ:

$$S(\tau) = \sum_{r \in R_\tau} logP(r) - \beta \cdot \sum_{nt \in NT_\tau} H(nt) \quad (12)$$

where R_τ is the sequence of rules employed during the derivation of τ and NT_τ is the set of all the nonterminals also used during the derivation of τ.

Finally, it is still necessary a complete definition of how entropies interact with probabilities. That is, we expect that the parser will mainly pursue high probability paths and only recur to entropies to disambiguate when the analyses show similar probabilities. How small are the weights of entropies and how they evolve during processing is something that still needs to be empirically defined.

4 Experiments

The parser with which the experiments were performed is an augmented version of the system developed by Roger Levy (2008), which is an implementation of the Stolcke parser (Stolcke, 1995). The latter is itself an extension of the Earley Parsing Algorithm (Earley, 1970) that calculates incrementally prefix probabilities.

The grammar was extracted from the sections 02–22 of the Penn Treebank (Marcus et al., 1993). Applying transformations on the trees would also change the entropy values, adding another variable to the model. Because of that, the only transformations performed were the removal of functional annotations and a preprocessing using the software by David Vadas (2010).

4.1 Formulation Refinement

As mentioned previously, the second part of our model's formulation still needed to be refined in order to arrive to an optimal trade off between the two addends. To do so, we used the section 24 of the Penn Treebank as development set.

First, the best 50 parses according to the traditional PCFG parse probability score were retrieved. For each of the retrieved trees, f-scores were associated using Evalb and the gold standard trees. Then we proceeded to re-rank them according to different variations of our model. Several functions were tested in order to combine the entropies with the probabilities.

We found that the contribution of the entropy of each nonterminal had to be in function of its location within the tree. More precisely, the contribution $H_L(nt)$ of each nonterminal *nt* varied according to the distance between the location of the *nt* and the beginning of the tree according to a preorder representation. For example, if we have the following tree in preorder representation (terminal symbols were removed for readability):

$$(ROOT(S(NP(DT)(NNS))(ADVP(RB))(VP(VBD))(.)))$$

the distance of "NP" in the tree to the root node "ROOT" is 1 because there is one nonterminal symbol between "NP" and "ROOT". Likewise, the distance to "ADVP" is 4 for the same reason. Thus, if we define the distance of *nt* to the beginning of the tree or root, a localized entropy contribution of each nonterminal H_L can be defined as:

$$H_L(nt) = \frac{H(nt)}{Distance(root,nt)^{\alpha}} \tag{13}$$

where $H(nt)$ is the entropy of the nonterminal *nt* according to the grammar, and α is another parameter that needed to be found and that denotes the polynomial decrease of entropy's relevance according to its location.

In order to translate this into the actual parser, it was necessary to find an approximation of the distance measure that could be calculated during parsing when the complete tree is not yet known. One approximation was the following:

$$Distance(root,nt) = 2i + (\mid x \mid - j) \tag{14}$$

where (i,j) are the coordinates of the nonterminal in the Earley parsing chart and $\mid x \mid$ is the size of the sentence. This formula takes advantage of the chart which could be seen as a tree where the root is in position $(0, |x|)$ and the leaves are on the diagonal.

Having a measure of distance, β and α were tuned using grid search, finding that 0.6 and 2 were good values for β and α respectively. After this, we have all the elements to get a final formulation by plugging H_L into our model:

$$S(\tau) = \sum_{r \in R(\tau)} logP(r) - 0.6 \cdot \sum_{nt \in NT_\tau} H_L(nt) \qquad (15)$$

4.2 Results

Having generated the grammar and tuned the parser, experiments were conducted on the section 23 of the Penn Treebank (2416 sentences). For each sentence the parser was fed with the gold POS-tags in order to avoid tokenization errors and errors concerning words that did not appear in the training set.

The Baseline was a parser that retrieved the best analyses according only to probabilities. The resulting trees were compared to the gold standard using Evalb, obtaining the results shown in Table 1.

Tab. 1: Comparison of the output of Evalb for the baseline and our model.

	Baseline	P-H	Δ
Number of Sentences	2416	2416	
Bracketing Recall	70.39	**71.41**	+1.02
Bracketing Precision	75.73	**76.94**	+1.21
Bracketing F-Score	72.96	**74.07**	+1.11
Complete Match	9.73	10.10	+0.37
Average Crossing	2.89	2.67	-0.22
No Crossing	33.57	35.64	+2.07
2 or less Crossing	59.27	63.16	+3.89

The first numeric column pertains to the Baseline, the second one corresponds to our system and the third one shows the difference between them. We see that although the improvement is modest, the new model outperforms the baseline in every field, suggesting overall benefits of the model.

In order to have a better characterization of what exactly the parser is doing better, another analysis was performed. In this case the output was converted to dependency trees using the software by Johansson and Nugues (2007) using options in order to emulate the conventions used in CoNLL Shared Task 2007.

After obtaining the dependency trees, the output was evaluated using the CoNLL-07 shared task evaluation script. This gives a more fine grained output that specifies among other information how the parser performed for each type of dependency relation; which can be seen in Table 2, where light gray highlighted rows correspond to dependencies where the new model outperformed the baseline and dark gray where the opposite occurred.

Tab. 2: Precision and Recall of Dependency Relations.

		P-H		Baseline		Δ	
Dep.	**gold**	**Rec**	**Prec**	**Rec**	**Prec**	**Rec**	**Prec**
ADV	4085	80.02	63.07	79.19	62.77	0.83	0.3
AMOD	980	54.69	54.75	53.98	54.31	0.71	0.44
CC	188	91.49	90.53	91.49	90.53	0	0
COORD	2795	76.89	76.1	76.03	75.33	0.86	0.77
DEP	1072	85.45	75.39	85.54	75.41	-0.09	-0.02
IOBJ	296	33.45	26.26	33.78	26.67	-0.33	-0.41
NMOD	19515	84.16	92.75	83.77	92.23	0.39	0.52
OBJ	3497	60.22	65.22	60.08	64.95	0.14	0.27
P	6870	99.62	99.53	99.62	99.53	0	0
PMOD	5574	82.11	82.66	81.13	81.71	0.98	0.95
PRD	671	78.39	79.22	78.54	79.13	-0.15	0.09
PRN	140	49.29	60	49.29	59.48	0	0.52
PRT	159	100	88.33	100	88.33	0	0
ROOT	2416	83.2	83.2	81.87	81.87	1.33	1.33
VC	1871	90.86	74.5	91.29	75.04	-0.43	-0.54
VMOD	6555	84.52	80.17	84.03	79.47	0.49	0.7

As we can see, the most frequent dependency relations like NMOD, VMOD, PMOD, etc. were handled better by the new model. Infrequent dependencies show no difference. It is only slightly infrequent dependencies (DEP, IOBJ, VC) that show some disadvantage, perhaps due to data sparsity. Because of lack of space we are unable to show all the output of the evaluation script, however, the rest of the results are very similar to Table 2: frequent structures were handled better, infrequent ones showed no difference, and only slightly infrequent ones show a disadvantage for the new model. Namely, this pattern arose for dependency relations across POS-tags and dependency relations plus attachment. Finally the new model gave a general improvement regarding the direction of the dependencies and regarding long distance dependencies.

4.3 Discussion

Addressing the formulation refinement, we see that the distance of each nonterminal to the root is an important factor. Indeed, it was only after considering this aspect, that some real improvement was obtained during development trials.

Its form entails a rapid decrease of the relevance of entropies as the sentence develops. One possible explanation resides on the predictive nature of the model. At the beginning of production it is possible that the speaker constructs the syntactic structure to be as concise as possible, making decisions that should lead the parser to quick and/or highly probable paths. As the speaker continues, the possible continuations become less and less as syntactic and semantic restrictions arise. At the end, it is possible that the speaker has very few options, so the only paths available are specifically those that match the semantic and syntactic representations already constructed.

Another factor that could partially explain this phenomenon is that during parsing, nodes nearer to the leaves are more determined. The very first level of parsing, POS-tagging, is a task that has a very high degree of accuracy; as the nodes go up on the tree, this determinacy is less and therefore the need for non-local accounts arises. The distance measure here used precisely gives less relevance to entropies that reside at lower levels of the tree.

It is worth mentioning that even though the relevance of entropy lowers as the analysis continues, it never reaches zero. That is, at the beginning of the sentence the weight of entropy can possibly override that of probabilities, making the parser follow paths that are not necessarily the most probable. This power of overriding probabilities diminishes as the parser continues, but even though it can be very small, for structures with equal probabilities, the parser would still prefer the one with lowest entropy.

Regarding the results of the experiments, we can see a modest but general improvement; only few structures are not handled better, and those happen to be rather infrequent.

The new model performs better especially for structures that are source of attachment ambiguities, such as coordination and long distance dependencies. This further supports the model as a means for preferring shorter and simpler structures when it comes to decide between structures of equal probability.

We should not expect every structure to seek a quick finalization, one could argue that there must be structures that actually mark that the sentence is going to be longer or shorter. It seems reasonable that the relevance of entropy depends also on lexical, syntactic or semantic information. A more complete model with this information, as well as experimentation with other languages and grammars should give more precise ways of using entropies during parsing.

5 Conclusion

We proposed a framework that makes use of entropies as a way to predict/weight syntactic trees. Models within this framework are argued to be pursuing three rational goals: a)retrieving the most probable analyses according to experience, b) being quick in order to cope with communication time constraints, and c)retrieving analyses minimizing cognitive costs.

With this, we implemented a model that provides a definition of syntactic parsing at the computational level, ranking parses according to a trade-off between probability and entropy. After testing, we concluded that it is indeed beneficial to use probabilities and entropies to rank syntactic trees.

Bibliography

John Anderson. The place of cognitive architectures in rational analysis. In K. VanLehn, editor, *Architectures for Cognition*. Lawrence Erlbaum, Hillsdale, NJ, 1991.

Nicholas Chater, Matthew Crocker, and Martin Pickering. The rational analysis of inquiry: The case of parsing. In Chater and Oaksford, editors, *Rational Models of Cognition*. Oxford University Press, Oxford, 1998.

Jay Earley. An efficient context-free parsing algorithm. *Communications of the ACM*, 13(2): 94–102, 1970.

Ulf Grenander. *Syntax-controlled probabilities*. Division of Applied Mathematics, Brown University, 1967.

John Hale. A probabilistic earley parser as a psycholinguistic model. In *Proceedings of the second meeting of the North American Chapter of the Association for Computational Linguistics on Language technologies*, pages 1–8. Association for Computational Linguistics, 2001.

John Hale. Uncertainty about the rest of the sentence. *Cognitive Science*, 30(4):643–672, 2006.

John Hale. What a rational parser would do. *Cognitive Science*, 35(3):399–443, 2011. URL http://dx.doi.org/10.1111/j.1551-6709.2010.01145.x.

TF Jaeger and Roger P Levy. Speakers optimize information density through syntactic reduction. In *Advances in neural information processing systems*, pages 849–856, 2006.

Richard Johansson and Pierre Nugues. Extended constituent-to-dependency conversion for English. In *Proceedings of NODALIDA 2007*, pages 105–112, Tartu, Estonia, May 25–26 2007.

Roger Levy. Expectation-based syntactic comprehension. *Cognition*, 106(3):1126–1177, 2008.

Mitchell P Marcus, Mary Ann Marcinkiewicz, and Beatrice Santorini. Building a large annotated corpus of english: The penn treebank. *Computational linguistics*, 19(2):313–330, 1993.

Brian Roark. Expected surprisal and entropy. *Oregon Health & Science University, Tech. Rep*, 2011.

Claude E. Shannon. A mathematical theory of communication. *The Bell System Technical Journal*, 27:379–423, 623–656, July, October 1948.
Andreas Stolcke. An efficient probabilistic context-free parsing algorithm that computes prefix probabilities. *Computational linguistics*, 21(2):165–201, 1995.
David Vadas. Statistical parsing of noun phrase structure. 2010.

FSC
www.fsc.org
MIX
Papier | Fördert
gute Waldnutzung
FSC® C083411

Kolibri 360 GmbH
Ferdinand-Jühlke-Straße 7
99095 Erfurt, Deutschland
produktsicherheit@kolibri360.de